GOLD MINING! COME ALONG FOR THE RIDE

Tom "Lefty" Leftwich

PublishAmerica
Baltimore

First printing

ISBN: 1-60610-217-6
PUBLISHED BY PUBLISHAMERICA, LLLP
www.publishamerica.com
Baltimore

Printed in the United States of America

This book is dedicated to my best friend and partner for life,
my wife Fran.

Acknowledgments

I wish to acknowledge and thank my many gold mining friends and the ICMJ Prospecting and Mining Journal for their encouragement and support. Particularly I want to thank Gail Scott my computer mentor, and the publisher "Publish America" for their fantastic aide and program which made this work possible.

Table of Contents

Chapter 1
Beginning of a Gold Miner

Where does a gold miner's life start? Looking back, I reckon that maybe fate has something to do with it. I was in the eight grade in Lafayette Jr. High School No.79, Baltimore Maryland and my teacher's name was Miss Fox. Why I should remember this and the details; I haven't a clue. I was given an assignment to write a book report on some western style literature written by a feller by the name of Bret Harte. His series of short stories was kinda right up my alley because I was a number one western movie fan and a potential school drop out at that time. Some of that recommended ol'English literature, that I had been forced to stay awake through, was so boring; it would bring tears to a bulldog. I can't tell you of another book report that I ever did. I'm sure there were many but, my memory comes up short on any others.

Well, now ol'Bret told some good stories and I really enjoyed all of them. "The Outcasts of Poker Flats" sorta grabbed my attention and although I liked his other mining camp stories, I selected this one maybe because it was shorter than most and I could whip out a quick report. My memory fades out at this point, and I can't tell you if I even got a passing grade on my composition. Did this story inspire me to become a gold miner? Of course not! I don't remember there being a whole lot of anything in his stories that got into a lot of detail concerning gold mining. What with a bunch of teenage girls around, I reckon my interests were otherwise occupied.

At this point, I've got to take a leap into the future because my next connection with gold mining jumps to 1960 when I purchased a home in Rosamond, California next door to the Tropico Gold mine. How this all came about is a tale in it's self but I still to this point had no interest in gold mining. Life goes on for all of us, I met a fantastic girl Fran, got married, started a family, and in 1965 Glen and Doreen Settle, owners of the Tropico Gold Mine; asked me to help them in putting on a World's Championship Gold Panning Contest. I and Fran, with family; were involved with the Rosamond Rider's Horse Club,at that time; and it was natural for us to become involved in anything our neighbors were doing. I was asked to handle the Stake Claim Race for the kids and my pretty pinto mare "Gypsy" rearing up to start the race was a colorful beginning for the event. Our kids were in the contest in later years and enjoyed it very much. I still had no interest in gold mining.

Along about 1970, I decided to quit bull riding in Rodeo's (I wasn't exactly setting records anyway) and take up a tamer hobby that gave me more time to earn a living and less time, laid up; healing. I picked up a guitar, for seven dollars and following the long suffering ordeal that Fran had to endure why "Look-a here Mama" I'm a country singing cowboy! Well I ain't braggin, but some folks did like my entertaining and I went on to become a half way descent radio gospel singer. During the 1970's I was committed to every entertaining and Master of Ceremonies function that took place in our community and became more involved with the World's Championship Gold Panning. As time went on,I became M.C. and entertainer for this event. My wife, Fran sorta jumped in with both feet and became a world class speed gold panner finishing third one year. As for myself, I still had no interest in gold mining.

"Look-a-here Mama, A Country Singing Cowboy"

What's Gold Panning? Well, the 49'ers coming to California for the gold rush had no knowledge of how to separate dirt and gravel from the gold. Most of them learned in two days to become a good panner. They took a skillet or pan with sloping sides, filled it with dirt and a couple of buck shot and washed the dirt out until they could retrieve the buck shot. When they were able to do this with some degree of success, they were plenty good enough to go gold mining. The Gold Speed Panning Contest, is a carry over of this original technique used by the 49'ers although the methods have been modified for speed competition.

Thinking back, I reckon it had to be around 1976 at the World's Gold Panning Contest that I met Elton Rodman owner of the Roaring Camp Mining Company Resort located in northern California. I was sitting there playing "old country" to entertain the gold panners at Tropico when Elton invited me to come to Roaring Camp for his annual 49'er Gold Panning Championship Day. He sorta felt that my kind of music was what he needed for his customers and he'd give me a free vacation with accommodations if I would come. At that time I was still gainfully employed and far too involved with Rosamond activities, President of the School Board,Horse club, Little League Manager, Gospel Singer for Local Radio and Director of the Edwards AFB Mission Control Center(my day job), to consider adding anything else. My cup done runneth over and then some!!

The World's Championship Contest moved to Knott's Berry Farm and then to Reno, Nev. And each time we ran into Elton and each time he invited me again and each time I had to decline. Then I was entertaining and MC' ing at gold panning contests all over California, when I ran into him at one, out here on the desert at California City.

Well I guess he had worn me down, cause I looked at my schedule and accepted his invitation. The fact that friends of mine from Ahwahnee won his award of a free week at Roaring Camp and I would be there with them, was a deciding factor.

Now right here!! is where life gets back at you and you realize that maybe you aren't as much in charge of your destiny as you think you are! There on the entrance to the gate going into Roaring Camp was a specter from my past. A startling jolt from my teenage years and the eight grade. The burned in words from one of Bret Harte's stories, "The Luck of Roaring Camp"!! I was floored! Later to find out that this canyon was one he had visited and served as the inspiration for his story. You know a feller has to kinda pick himself up after this kind of an awesome happening. All these years to have that book report filed away just waiting to wake me up to the mysteries this ol' life holds for all of us. I had to feel that somehow fate had set me on this course from the very beginning. A full continent and 33 years away! Man is this some kind of Science Fiction,or what? Roaring Camp had got it's name from Bret Harte's story! Here I sit, mouth hanging open and wondering if I'm really awake.

"Sign at the gate of Roaring Camp"

I had to weather the shock and see what was in store for me. Well, this turned out to be the greatest week I ever spent at Roaring Camp, I fished, I swam,I butt walloped down the rapids with my two young daughters. Fran and I enjoyed the company of good friend "The General" from Ahwahnee, California. I even helped him dredge(an under water vacuum cleaner like thing) for gold a little (very little) with daughter's Loi and Loreea. It was a great week because we could care less about gold and didn't get involved in that so called Recreational Gold Mining (better referred to as HARD WORK!!!) What a wonderful vacation, I wrote a song or two, played a little music, met lots of great folks, made some very beautiful memories. The girls singing at midnight under the camp Manager's window (a frog on a log in the bottom of the sea) then runnin off and leavin him holdin the bag on a Snipe hunt. Yes, we made plans to go back the following year!!

Chapter 2

The Language of the Gold Miner

I reckon that as a Green Horn at Roaring Camp, I was about as dumb as an oak post when it came to talking Gold Miner Lingo. Reflecting back on my own experiences, I'd say at this time in order to understand what I'm talking about in the rest of this story, a feller or a lady should know what some of this terminology means, so I'm gonna interrupt this thrilling episode to throw a little education your way. Most folks don't know a lot about gold and I gotta tell you that you sure aren't alone. There just ain't no experts when it comes to gold and I don't claim to be no different. It's just that my circumstances have allowed me to become a little more familiar with it, than most folks.

Well, to begin with gold is 19 times heavier than an equal volume of water, so to illustrate this a pint of water weighs 1 pound, so conversely a pint of gold would weigh around 19 pounds. Now I know this is not exactly correct because gold is weighed in Troy ounces dating back to the Roman Empire, and of course there is a difference; but if I get into a pint of gold, I sure ain't gonna worry about a little difference in Troy versus English weight. As a point of interest 20 pennyweight (dwt.) equals one ounce Troy and 12 oz. Troy equals 1 pound. (That's close enough for most purposes)

Gold is found in two major conditions(what someone else owns and that what you're looking for—just kidding!). Sorry, but I don't want you to get bogged down with this learning process. The first type of gold, is described as Lode gold wherein the gold is still trapped in it's original formation (fused inside rock carrier material). This carrier material was extruded from the center of the earth during it's formation and gold was formed at that time being carried to the earth surface within this molten carrier material. The location of these carrier seams has resulted in the underground mining techniques and development of mile plus deep gold mines. This condition requires crushing the rock like material to free the gold for processing. The methods used for this are somewhat determined by the size and character of the gold. Most generally the Lode seam carrying gold is mined from large open pits or underground tunnels and processed at a Mill site designed for this specific purpose. To this date, the majority of recovered gold since 1849 is from lode gold and underground mining.

The second and most commonly known is called Placer gold. This is gold that has been released from it's original earth formation carrier material by aging and weathering of the earth surface. Prior to 1849, most of the gold accumulated in the world was from placer or free gold operations. This type of gold can be found almost anywhere on the common continents as scientist have discovered that ice age glacial movement spread river deposits in many areas where gold would not commonly be found. The rich deposit (Mother Lode) area of California encompasses the old Highway 49 Gold mining towns from Mariposa to north of Placerville. This is the major area worked during the original 1849 Gold Boom in California. My stomping grounds(Roaring Camp) is just up the road from the original wild, wooley, and wide open Jackson Town and slap dab in the heart of the Mother Lode.

The heavy weight characteristics of Placer gold versus all other common minerals causes it to migrate by gravity to the lowest point possible (like River beds), and this characteristic makes it relatively easy to separate from base material. The most common form of separation was in use by the early Egyptians using a rock made sluice box which employed stones as riffles to trap the gold. This was followed by wooden devices and todays metal sluice boxes all of which utilize this same or similar technique to separate gold from it's accompanying material (gravel and sand or soil base). The sluice box would have to be described as a rectangular trough possessing cross bars or riffles over which a flow of water would be directed to wash base dirt and gravel in order to permit heavier material such as gold to become trapped behind or in front of the riffles. Normally the sluice box is tipped at an angle downward (5 to 10 degrees) to get proper water flow. The water in today's boxes is introduced through a disperser or spray bar at the top of the box. In ancient times, it was hand dipped and supplied to the upper box in a bucketing manner.

The ol'timer's, referred to earlier; came to California during the 49'er Gold Rush and knew absolutely nothing of mining techniques, but within two day's they were panning out gold. A technique which permits a washing activity to allow the heavies to settle and separate from base material. Dry washing was another technique used for separation of material in a dry or desert location.

So, I say it doesn't take a rocket scientist to learn mining techniques and separate material to get the gold. But at the same time, I do say that it takes years of experience and know how to make an educated assessment as to where one would most likely find Gold.

OK, so let's cover the equipment used by today's Gold miner. The Gold Pan, there are a thousand different verities and they will all work to a degree. The operator is the most important factor in recognizing the action of gold and the characteristics of the gold pan in use. Any good gold panner can use a 10 inch skillet and get the gold! The 49'ers cooked in it, ate out of it, burned out the grease and panned out their gold. Personally, I prefer the Roaring Camp Pan, because it's light weight aerospace plastic, almost indestructible, has under cut Chinese riffles, and a wide flat bottom for separation classification. Once again, I repeat; to become an excellent gold panner, all one needs is a gold pan, some buck shots, some dirt and a tub of water. Practice will do the rest. Wash out those buck shots and you're just as good as you have to be to get the gold. (Note: Chinese Riffles were first developed by the Chinese using a wooden type bowl with cut in riffles on the side walls to prevent the loss of fine gold during the washing process. This improvement has been adapted by all modern Gold Pan manufacturers.) As a point of interest a gold pan is excellent for cleaning fish, just in case your gold prospecting is not too good.

The sluice box is the basic classifier used for a larger volume operation. Todays boxes can be set up in the river and held in place by placing a large flat rock on the sluice way about half way down. The angle and water flow is always a concern for the operator, so let's do this. Drop a handful of pebbles in the upper end of the box and all or most, of the one inch variety should be washed out by the water action. This is usually a fairly good rule of thumb to use. The proper water action is helpful, but even when the supply is a little too much, gold will tend to stay in the box. Most manufacturers will provide operator instructions. Regardless of type of sluice box, panning out the contents is still required to get the gold.

OK, so what's this High Banking thing all about? High Banking is a term used when you're working out of the river and must transport gold bearing material to your sluice box. You must provide water for this operation, so a water pump with hoses for delivery is required. Generally, I will set up as close as possible to my mining area, to reduce the material transport requirement. A settling pond maintained at the end of your sluice box will considerably reduce the amount of silt that makes it's way back to the river. The environmental folks will appreciate this and the Fish and Game will insist on it. A gold miner should always be aware of doing things in a manner that protects the environment.

You'll hear some folks talk of a Rocker Box, well this is a sluice with rockers that permit the operator to rock the box sideways in order to get more washing action. They are usually small and require the operator to dip and pour the water required for operation. I'm sorry, but I'm not a fan of this technique. Some one else is welcome to my share.

Let's get into dredging, now here is an instrument that really moves the dirt! The operation employs a floating sluice box with a high pressure water pump, an air breathing compressor and power jet venturi which provides a vacuum action for sucking material from the river bottom for delivery to the floating sluice box. The compressor provides air for the underwater operator and vast volumes of material can be ran in a day. Considering time of operation and effectiveness, this system ranks first. The biggest limitation is boulders and they're always an impediment to any operation. The biggest drawbacks are; cost, complexity of operation and mining time required for benefit results. I always recommend this operation for seasoned miners that have already learned the basics from simpler operations.

OK, so I over looked crevicing but, I'm gonna cover that in some detail later on. Right now, let's say that lot's of folks enjoy this gold mining activity, because it appeals to people like myself that try to stay clear of high cost and physical work. Crevicing is the art of finding and cleaning out cracks and faults in bed rock that tend to collect gold particles and nuggets. This is an activity that can be conducted in the river bed underwater(called sniping) or on the shoreline (crevicing).

Well, you've made it through the familiarization class dealing with gold mining and by now you have a layman's idea of what kind of processes you're gonna run into, so I'm not going to be-labor you with a lot of explanation from here on out. Besides that, you're probably smarter than I am, by not being a gold miner. They do say, that a gold miner has got a screw loose somewhere anyway! But, in my case; Hooray for that loose screw, I've enjoyed every minute of the trip!

Chapter 3

The Wiss Bangr

The characters are real and only the language has been changed to protect the innocent! Following my introduction to Roaring Camp,I had evidently been slightly bitten by the gold bug, because when the General invited me to come visit him on the Fresno Rive and do a little gold mining; I immediately jumped at the chance.

By way of introduction Mr. Marshall Lee Sherman (nickname "General") and the fact that he was one dear individual, a friend whom I shall never forget. Here's a guy that started his gold mining career at 16 when he built his first sluice box, out of wood, of course. Our affiliation and involvement in gold mining was filled with some very exciting and entertaining times. We deeply enjoyed the pleasure of each other's company and he wanted me to go dredging with him. Now you gotta understand,I didn't know beans about dredging or anything else about gold mining at that time. Naturally, he explained all the wonderful advantages of dredging and led me to the slaughter.

This was July 82, but the river was co-o-old and I didn't have a wet suit(a rubber or vinyl suit such as surfers use). The General had outfitted me with a 2 inch dredge with a 20 foot suction hose. Now this thing woulda probably worked great for someone who knew what he was doin,. But me,I spent two minutes dredging and five minutes clearing the plug ups. The water was freezing cold and I'm thinking this sure ain't something that I couldn't live without. Well lunch time finally got there and just in time before I became a permanent blue.

Here comes the General and I lay my story on him as to how if this is all there is to gold mining, I jest ain't interested. Is there some other way to git some gold without freezing your butt off. He tells me he's got a Doodle Bug home in his basement. That is some kind of a high banker thing that you shovel dirt into. Well I'm game for anything, so after lunch, we set this thing up using the motor off the dredge to pump water to the sluice box. What's a Doodle bug sluice Box? Well, it's a long trough with cross bars called riffles that catch any gold that might be washed down against them. The water was pumped in from the top of the box and I shoveled dirt and gravel into the upper end. This action permitted the water to wash out and free any gold which would become trapped by the cross riffles. The riffle bars were removable and you could dump the contents into a gold pan for panning out the gold.

Next thing I says is where do I dig, you know there's about two hundred yards of ground in sight. He says, "Well you see that crevice coming down that wall of bed rock cliff, see if you can find it down here on the river shore and dig there." I can see he wants to go back to his dredging down river and didn't want to waste his whole day learning some rookie everything, so I kinda take it from there. I looked that shore line over and finally found the blame thing coming into the river about fifty yards upstream. So I set up this Doodle Bug thing and start shoveling out the crevice. Remember it's July and I ain't a hand at manual; labor (that's what my wife has been led to believe—at least I've been working on it) so anyway I figure after about half an hour I'm ready to find some other way to "dig fer gold."

Well I go back to that stop and go two inch dredge to clean out the crevice as far as it will go up the bank, that kinda seemed like the thing to do, seeing as how it sucked up all there was until it run outta water. Now I'm stuck, there's this shovel looking me in the face and a crevice there that looks uninviting, Now give a lazy man (me) a hard job and I might look for an easier way to do things. I figured if I could get water to that crevice,I should be able to suck that dirt out. That Doodle Bug thing I'ed already modified so the suction hose would dump into it's sluice box. I took the legs off so it would be close to the ground. This avoided the lift and made it work without too many plug ups when I was using it as a dredge. Next I look at this eductor thing(suction nozzle) and I figure the water is right there near the suction end, and if I had a valve or something, why I could fill up that crevice. Inspiration is a major tool of the lazy afflicted, so I laid the suction nozzel in the crevice and went back to the sluice box, blocked the delivery end of the suction hose with my hand, and there she was, water in the crevice which I immediately sucked up with a bunch of material. I ran back to the box and repeated the operation. After about an hour of this so called easy way to dig gold I'd just developed, I'm working myself silly, but boy that crevice for about ten feet is as clean as a bathtub.

Here comes the General once again, and this time, he stopped in his tracks and says, "What the heck kinda Wizz Bang operation you got goin on here?" (As a Note; His language was somewhat different than as written.) I said General I really got it made, look, at this and I showed him the operation. He was super impressed, but the big thing was in the sluice box, the first riffle shows gold all the way across. "God!,"he says, "I haven't seen this much gold in three days of dredging. Which end of this thing do you want,I'll take the other." We worked the rest of the day that way. I can't say how much gold we got, I wasn't much into that dwt stuff at that time, but it sure looked good in the bottle. In reflection, I'd say we had close to 2 penny weight(1/10 of an ounce) of gold.

Now for years in my day job at Edwards AFB, I had passed myself off as an engineer, so that night we sat down and designed the first Wizz Bang figuring water head pressure,delivery, vacuum etc.

In succeeding weeks, tests were run using different setups, but the two inch was most efficient.

Our success had been so phenomenal that maybe an attempt to patent was in order. I had to write an abstract thing to apply, and the official name had to describe the unit S-o-o-o it became the Water Induced Surface Sluicing of Bedrock Allowing Normal Gold Removal (WISS BANGR) named of course after what the General had called it. During the second week of August 1983, the first unit was demonstrated to BLM and Fish and Game folks from Fresno.(We actually found some gold during the demo) They said it was a neat recreational gold mining implement and used properly with a settling pond was legal for use with a permit. I asked if I could get a statement and signature to that effect. No,my friend, that turns out to require an act of congress which necessitates generation of the usual railroad car load of documents and a majority of electoral votes but we would suggest you get a dredging permit if you wish to continue using it. Well, the Wiss Bangr was born and the patent application fell through for what ever reason, and though the General and I never received any monetary consideration for it's invention,through out the following years; we found lots of gold with this unit.

As most folks know, Gold Mining is tough enough, and I'd like folks to enjoy this unit as I have,so I've included it's conceptual design for public use. When constructed properly and used in a legal manner, it does a great job.

The illustration and design, below; has been modified for back packing and now utilizes the 2 1/2 hp Honda motor with the Keene high delivery pump. A header box sluice is necessary for best operation as a large amount of air is present with the material and requires the stop and go action of the original header box designed by "Grand Pa" Keene. The flusher fire nozzle is a twist to shut off fogger nozzle used by the fire dept. A jet nozzle will blow your gold away and remember you're washing bedrock, not washing down a mountain.

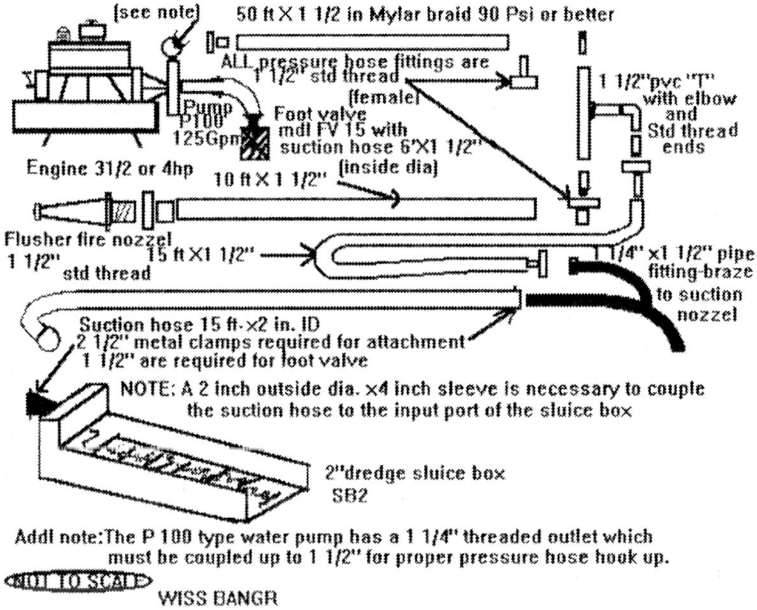

"Sketch Design of the Wiss Bangr"

How do you use this thing? Locate some shallow bed rock and wash it off into a depression or hole and use the suction nozzle to vacuum up the contents for delivery to the sluice box. Use the flusher nozzle to provide the water and washing action. It's a low cost portable unit that will work your butt off and get you some very nice gold. Always provide a settling pond off the end of the sluice box to prevent silt getting into the water supply (a la Fish and Game requirement). Most folks that use one, want one and you'll find them scattered out in Europe, Canada, Mexico and the United States. I don't think they've made it to Australia yet, but recently I did see an Aussie at Camp, drawing up a design for one.

Chapter 4
The China Camp Caves

The year,I think was 82 and once again we had come to Roaring Camp to enjoy a brief vacation and try out the newly invented Wiss Bangr. My first attempt at using it was at the John's Hole diggings above camp. Lord was this ever a mistake. For a while,I thought I was gonna get shot by everyone in camp for daring to use a new machine on John's Hole in such a manner that I might demonstrate a much easier way to git the gold. I think their biggest objection was that I had the only one of it's kind and they couldn't get one of their own. Anyway, I had to go to China Camp, if I wanted to use it; because it was causing too much trouble in camp.

So, off to the China Camp caves I go to look for a glory hole. The "General" was in camp with us that year and he had his own version of a Wiss Bangr,so we both set up in that area. There was a lot of fine gold,but not much over the flake size and we had to process a lot of dirt to get more than just color. After Fran and I had dug a hole in the limestone bedrock you could lose a Volkswagon in, and found almost enough picker size fine gold to cover the bottom of a 2 ounce bottle; we decided to move to a more attractive location that might produce better. The fine gold was really plentiful, but like everyone else; we were after that big stuff!

There was a large outcrop of bedrock just upriver at the beginning of the beach area in front of the China Camp caves. I really wasn't what you'd call a gold miner at that time, but I figured the river running high would deposit down stream just behind that out crop. Right on!! I did much better in this location, even captured a flat piece of gold sliding down the bedrock wall. The problems I encountered during this operation were two fold. Number one,on Tuesday morning, Norman hiked down river with a group of gold miners from camp and my first indication of this was when I hear an exclamation(not too nice) from above me and the 6 to 8ft. hole I've dug. It so happened my digging was slap dab in the middle of Norman's hiking trail taking it out completely and requiring him and his group to wade the river to get around. As a note;,I had to fill the hole up when I moved on but before I did, I was visited by Norman almost every day. He loved that Wiss Banging and I couldn't keep him outta the hole.

A second interruption was brought on by my equipment. We were working close to shore so I floated my motor/pump on an inner tube at the edge of the river. This worked great until while I was busily engaged mining, my motor died and I jumped outta the hole just in time to see my motor sinking slowly to the bottom of the river!! That inner tube done sprung a leak and down she goes. Now this was a major calamity, I figured good bye motor and goodbye mining operation. A hot motor sinking into the cold water didn't bode well for valves and a warped head. I just knew it was a goner, but the good ol' General had other ideas. We pulled her out, dumped all the liquids and he opened up the carburetor to allow everything to dry. Within a couple of hours, that ol' baby was running like a top and I was back to gold mining. Our two weeks, though; had come to an end and it was time to pack her in and make plans for next year.

At this point, I sorta ran into a slight problem. We had two five gallon buckets of black sand and they had to be carried to the top of Cardiac Ridge(so named for it's vertical steep climb characteristics). Well,I had to float our equipment down river to the

Fall's landing, so Fran was elected to carry the Sand. Thankfully, these buckets were not full, but they weighed in the neighborhood of 70 pounds each. I told her that, no problem; why, being of equal weight, they would balance her out on the way up the hill. She wasn't impressed at all!! Had she been a little, closer, I might have got hit with one.

That winter, I used a slow sluice(low water flow) to reduce this concentrate down to about 10 pounds and had a good friend assay my black sand. He only took 4 ounces and did a furnace assay. I still have his report(1100 ounces per ton). He wanted my whole 10 pounds, said he could make me a beautiful button of gold from what I had left. Now,I think back,all these years I've been looking for "big yeller" and ignored the Bonanza easily available in little gold. I think some of my future efforts are gonna have to deal with how to capture and classify all that little gold. Those big nuggets are really nice to have, but all the mining folks know that ninety five percent of all the gold is of the fine type. A gold miner should maybe get his priorities straight but, the quest for that big Bonanza Nugget goes on anyway. There's just something about finding a nice piece of gold or a nugget that kinda pales all that little gold. If a feller was making a living out of mining then fine gold would definitely be a major player but for the ol' lazy sunshine prospector like myself; fine gold sells pretty easy and is gone pretty quick. A big ol' nugget lasts a lifetime!

Chapter 5
The Double-Barreled Wiss Bangr

They do say, you give a lazy man a hard job, and he'll figure an easy way to do it. Well, I decided that my 5HP Briggs and Stratton engine could do more work than I was gittin outta it. Everything pointed to"let's get more gold," I decided to run two Wiss Bangs with the same motor. According to my figures B&S could handle it. So, in anticipation of this effort, I put together the necessary hoses, splitters, suction nozzles, sluice boxes etc. and headed for the Fresno River. The anticipated acceptance test procedure was a great success. It now appeared that I could float the motor in the middle of the stream and run hoses in both directions to the shore line and operate on both sides of the river at the same time. Our test on the Fresno River fully illustrated that a double barreled Wiss Bangr would work. The only problem I could foresee was, how do I get all this conglomerate of junk upriver at Roaring Camp. Hand carrying all this stuff was possible,but there ought to be an easier way.

It turned out there was an easier way.

I built a float using a 17 inch tube on an aluminum frame and used plastic milk creates to house all the parts and pieces. The motor, I mounted in the middle of the float so it would balance properly when running. Now at that time, there was no road to the caves at China Camp, so you had to hand carry everything down Cardiac Ridge trail to get to the river. Packing out up that trail is where it got it's name, talk about steep; I've climbed walls that were easier.

To get back to mining, we packed everything to the river and there assembled the flotilla which I launched as the USS Mockaloomie. I put the foot valve on the pump/motor unit, fired up the motor to about half speed, pushed off into the river floating behind and using the spraying water as a propellant, I motored that whole mess up river right to my planned mining area. Boy this was great!. I had everyone on shore urging me on. I did develop one slight problem just before reaching my destination. A plastic oil bottle fell over against the muffler and I filled the whole canyon with blue smoke. Other than that it was a great ride floating behind my powered surf board and guiding it between the river boulders. I guess it made a great show for folks on shore.

Fran and I then assembled this new version unit and went to gold mining. The Double Barreled Wiss Bang was a great hit with everyone, especially me, because Fran now; to get my attention for help,she just sprayed me from across river(instead of the enjoyment she normally got out of throwing rocks at me—kidding?). Norman came around that afternoon and after one look exclaimed "My God!! What are you gonna think of next??" Well, I told him that with a larger pump I could run four Wiss Bangs at the same time. "Don't even think about it!! says Fran, enough is enough. The gold was pretty good but the fun was great!!

We enjoyed the help of Fran's sister, Dee; and picked up a really nice piece of red quartz gold out of a crevice near the river. I think she enjoyed the experience,but I'm not sure. The work was a killer, just like all gold mining, but she did take home a nice piece of gold and memories of a lifetime.

Fran and I took our operation upriver to the side walls below what was called Jump Rock Crossing. The banks of the river in this area is all bed rock so I figured that if we could find a crevice it really could be very good. We found one but, there were many big rocks to move and at this time, we only had a big pry bar. I tried to work this crevice and Fran worked the down river side of a large boulder that separated us. Well, you have to understand that a double barreled Wiss Bangr requires a lot of hoses and working close together, all of these kinda lie close to and over each other. When a lady dredger came down stream snorkeling to see our operation, she was appalled at the confusing array of hoses and equipment. She wasn't confused though, by the nugget we showed her. What a great day this was for Fran and I, we had just found our first good sized nugget. You know, I do have a tendency to minimize the level of work involved, believe me standing upside down on your elbows to clean out a crevice is really work even if I do deny the charge. It was a great day for us,and a wonderful beginning to many great. years to come.

Chapter 6
September Island

It was September 1985 and I was mining at the forks in the river on a ledge island, later to be referred to as September Island. I was using my recently developed Wiss Bangr and finding a little gold. My wife, Fran was working in camp as (Camp Host) assistant, so most of this operation, I had to do alone. Norman(Camp Manager) was working across the river in a bedrock trough, and I stopped on my way up river to visit. Earlier that year, I had given him a Wiss Bangr that I had built up from some spare parts. He was definitely enjoying his operation and when I showed up,He told me he hadn't seen much gold. I checked out his diggings and right away I spotted pieces of gold on small ledges half way and all the way up on the side walls of the trough he was working in. I pointed these out to him and explained to him he couldn't spray the water so hard, it just blew all the gold out of his hole. He was more than a little surprised and was never again accused of over use of water.

Well on this particular day, Fran was working with me during her afternoon break and when quitting, we stopped by to check on Norman's success. He was doin pretty good, he'd found a bunch of small gold and one flat piece about the size of a nickel, definitely a nice piece of gold, probably about 1 ½ penny weight.

We enjoyed his good fortune and excitement and headed on back to camp. At the Jump Rock crossing we encountered a couple that had seen Norman's gold and excitedly exclaimed that it was the size of a "Quarter." By the time we got back to camp, we weren't sure we'd seen the same piece of gold, the excitement was running rampant and everyone was telling us it was the size of a "Silver Dollar." Sometimes I think gold miners are as bad or worse than fishermen.

The following Monday it was Fran's day off and we decided to put in a good days work because I'd found a couple of wheat size nuggets following a gravel trail on September Island. Now the day before, a number of vacationers had come into camp and been pretty excited about the success we were having with the Wiss Bangr. They all wanted to see it in action. We told them we'd be working down river the next day,so come on down we'll give you a first hand look see. At that time, there were only three in existence.

When the crowd arrived, there were seven or eight (some were afraid to cross the river and some fell in the river). It was a friendly bunch, all having a great time. I was busy, head down in a hole about 3 ½ ft deep, and when I looked up I couldn't throw a rock out. The hole was surrounded by people with video camera's. In reviewing the video's I can see I was just a little disgusted at this interruption to my gold mining. Well, I threw that hand sized rock up river and when I looked back down to where it had rested, there was a small depression in the bed rock and it was full of GOLD!!. Talk about excitement, there was so much hootin and hollerin goin on and I couldn't get down close enough to pick up the pieces except with a long hook billed tweezers. I took my suction nozzle and a low spray of water and was able to pick up all the fines. I took the rest out with tweezers. God! what an experience. I've never been able to duplicate that find of a lifetime, not to mention, with camera's rollin.

That morning, Fran and I quit work at ten o'clock so we could just go back to camp and enjoy. How could you possibly work with that kind of excitement, you just had to go share it with everyone.

Well, for us, at that time and at anytime, this was a lot of gold 22 ½ penny weight which included nuggets of 2 ½ and 3 ½ dwt. We worked the hole down, takin out gold all the way to river bottom, where high water, at some time past, blew out everything. There were no more Bonanza's waitin for us. It had been a great three weeks or so for us, because our rewards for diggin may have been a total of three ounces plus in gold, but thousands of pounds of excitement and great memories. Other great finds, we'd make later on, but none like Whizz Banging the Mockaloomie on September Island.

For lack of a better description, this Gold find took place on an island in September, which later reverted to September Island and the name stuck.

The following year, a miner came to camp on a day trip, and wanted to know how to get to September Island where some guy had found

"NINE" ounces of gold under a small rock. You know, I'd kinda like to find that island myself.

Chapter 7
Moon Rock

On many occasions, I had the opportunity to do a little crevicing at a location on the river called Moon Rock. It's a fairly large flat cratered black rock outcrop on the inside bend of the river that kinda resembles the moon surface. Every time I had occasion to do a little work there,I always got some gold. Now I do refer to doing a little work there and I do emphasize Little, because my natural inclination where work is concerned has always been in that direction. This place always fascinated me because I could almost smell the gold there. The black rock was some kinda hard though and it took lots of hammering to get any where s-o-o-o naturally,I didn't. I did bring my Whiss Bang down there though and set up to do a little gold gittin—WRONG!

When I started pumping water the whole basin started filling up. My sluice box was flooding me out. I had to relocate and place my box on the edge of the basin and work inward from there. Man,this was just about no good,because I couldn't git no where,but I did see enough gold to give me a ravenous appetite for more. There was virtually a hundred square yards of that place that I couldn't git to. I tried everything,but still got flooded out. Mother Nature in her wisdom had done a good job of protecting her gold from some vagrant Wiss Banger. I packed up my stuff, muttered a few unkind word's to the river in general, and sought me a new glory hole. But!!

I didn't forget!! A good'ol die hard dirty ol'gold miner don't just pass away a little adversity. I do intend to git even!!

That winter, I built me up a genuine Whiss Bangr of 1 ½ in size with a smaller sluice box and a 35 ft. suction hose. I coupled this rascal to my 5HP motor/pump and I was ready for Moon Rock!! The river was a little high the following spring, so I didn't git in as soon as I'd hoped so I bided my time doin a little exploratory work up river. This upriver look see did me well in later years because it pointed me towards some more gold bearing area's. I worked past Slaughter House Gulch(upriver from Camp) and found a spot at the base of the tallest pine tree that had a pretty good showing for the little work I did. I was just prospecting with my bag-of-tricks tote and a gold pan, but the color looked real promising. Unfortunately: It was time to git back to MOON ROCK!!

Now it always takes me a few trips to haul all my stuff down river and I usually haul that 40+ pound motor first, because if I did it the other way around, I'd probably talk myself into waitin the morrow for a motor trip. Anyway I'm headin down past the Marbel Slab area below camp and I see the settling hole by (what we call)Little Lone Rock is full of water. Hey, now this is interesting because I'm a figurin, I could pump that water up to the drop off of Marbel Slab Crossing and Whiss Bang the tail end where it empties into a straight rock trough. This was an area always under a little water with a lot of rough crevices and pot holes. An absolutely perfect spot to catch gold. Nobody ever creviced there because it was always murkey and dirtied up as soon as you got started. This is what the Wiss Bang was built for, a beautiful place to work. I didn't get any further that week. I put'er in right there.

No sooner said than here come the gold, man I couldn't believe the luck. Here I am blowin the gold outta cracks that have sat there for hundreds or thousands of years without so much as a howdy from passing gold miners. That dirty ol'murkey water done kept them out until here come a dirty ol'murkey Wiss Bangr. Yep! she really paid off, I still wasn't much into that dwt. stuff yet, but that bottle sure looked good. Norman come around bout that time with a bunch of hiker's he's taking down river to do some crevicing. He was astounded at the amount of gold I had laying on the rocks, all nice pretty finger pickers, beautiful stuff!! He just couldn't get over it said he'd walked over that spot a thousand times and stepped right on where I was working and never figured there was anything there. Them folk's didn't want to go, they figured that right there was the place to be, but without a Whiss Bang,no dice,you couldn't see anything. Well, I cleaned her out and finally got tired of beating the rocks. It was time for me to move on down to Moon Rock—That Glory Hole was a callin to me.

My set-up with the smaller Wiss Bangr was great, slower at movin dirt, but it really worked good. Delivered all my water to the outside edge of the basin, and allowed me to work the whole thing. Most folks woulda been anxious to clean fast, get what you could and get on with it. I probably would have too, except you see,I had all summer to work, so I took my good ol'sweet time workin on my knees (with knee Pads) and just cracked open all the crevices, washed the bed rock, and savored every finger picker I picked up. God! It was wonderful!! I spent the whole of ten days working Moon Rock, enjoyed every minute of it. Every day was a beautiful bottle day—Like I said, I wasn't in to that dwt. stuff yet and probably enjoyed it more for being that way.

You see, every little piece of gold you find is not a measure of it's value in (How much is it Worth??) but rather a feeling that brings an unbelievable excitement and pleasure knowing that you have just found a valuable piece of the formation of this earth and that YOU are the first one to find that particular piece. It is an almost indescribable feeling of joy and wondrous luck. It may be hard for me to describe it and hard for anyone to understand it. But when

YOU, find it,you'll know just exactly what I mean!! There's really nothing like it and this is where gold fever really starts. It's not the monetary value that keeps that gold fever going, it's the lure of the quest and the feeling of success that keeps the spirit churning always looking for that height of satisfaction and once found, eager for more! This is where the term "Gold Rush" really applies and gold fever is born. Fortunately for me, I'm a lazy feller that steers clear of hard labor, whenever possible, and since anything concerning gold mining entails work! Why, I just never acquired gold fever. I really relish though, and enjoy the "Gold Rush" that occurs with every find regardless of size or value.

On with the attack on Moon Rock, Yep that's what it was, A beatin and a hammerin and a pickin up the gold. One day towards the middle of my stay there, a vacationing family, came down river snipe-ing. This is a term used when you work under water cleaning crevices in bedrock. Some times you can use under water air from a floating compressor, or snorkel with fins or I've used a long tube with a sealed window in one end. It's just another way to enjoy a little recreational gold mining. This particular group had a floating compressor wet suits and weight belts. Man they were some serious worker's, did pretty good too. They had a couple of nice little BB sized nuggets and a bunch of flakes they'd found.

I had just relocated to a rib of black bedrock that jutted up from the base about two feet and it had a thumbnail size opening on the side wall near the top. It was jammed with little pebbles almost cemented in place. Well the gentleman that was head of this snipe-ing clan was kinda scoffing at my choice of location to work. He seemed to feel that the bottom area would be more productive and that I might be wastin my time on this high cleavity. Now,that hole was filled with very small red and black pebbles. In my experience these, were not today's common river deposits and I explained to him that at one time, this black rib coulda been the bottom of the river and that gold coulda been deposited in that little hole. His comments to his family and me, indicated his opinion otherwise.

Well his family had come up to witness the action and knowing what the chances were of finding gold in any crevice, promising or otherwise, I was a little irked that, his sage comments would probably be justified and my efforts wasted. I did something,I don't usually do, roughly chipped out the crevice and blasted the hole with a jet of water. GOLD!!! flew every where, all over the bed rock under that rib and down around my feet. God!1 I couldn't believe it, The whole family was shocked!! I backed them away from the area so I could find all the gold. I recovered nine finger pickers and small nuggets combined. The sad part is, I'm not sure that I got all of it. I had blasted it all over the tailing pile that I was creating!! Man,what a dumb thing to do!! The little gold,I know I didn't get, it washed into everywhere around. I had to settle for all I could find and promise myself never to get in a hurry again and try not to get upset at some one's criticism. I went back twice to that tailing pile threatening to run it all again. I didn't do it all but, I did get a couple more pickers from the outer edges.

The snipe-ing family, got all excited about all the gold, kept wanting to see the bottle which renewed their enthusiasm and they took off running to get back to looking for their own gold. I wrapped up my operation that week and packed her off for another adventure and "possible" Glory Hole upriver, at Jump Rock Crossing.

Chapter 8
Jump Rock Crossing

Jump Rock Crossing got it's name from just that. If you wanted to cross the river at that point,you had to jump the rocks. On some previous occasions, I had visited this place and found a little gold, nothing great but flakes, fine gold and once in a great while a nice finger picker. I had always figured that there had to be a deposit there somewhere because the river necked right at the crossing. All the gold found further down stream had to come thru here and I just knew that somewhere some of it had to get hung up, I could almost smell the stuff (again). Search as I did, there didn't appear to be a good unfound Glory Hole.

I sat up the trusty Wiss Bangr and dug—in. Things right away didn't look so good. I exploited around for a day or so and had just about decided to move on when ol'Richard,a miner of some minor repute, came to camp. Well, his stories and many claims to fame kinda enticed me to stick around awhile,so I sat to moving a few rocks and picked up a couple of flakes. This much encouragement was enough to get me curious about a big flat house sized rock that bridged a bunch of rocks beneath it. I decided to clean the bed rock under that big Mama and take pot luck. I hadn't been at it a day when I started uncovering a two foot whirl hole in bedrock directly underneath Big Mama.

Normally I didn't fool with whirl holes because I'd never found anything in them. I did think though,that just maybe this hole hadn't been visited lately, so I give her a shot. Right ON!! That hole was a good two feet deep and produced half a dozen nice pieces, probably more than a penny weight, if I had known what a penny weight was at that time. Hey!! This was wonderful, isn't it great that I had the fore sight and great mining skills (dumb luck!!) necessary to continue working in this area. (Fran really didn't have to put her comments in this story,you know) Anyway,at this point I'm congratulating myself and go over to visit ol'Richard.

Richard was pleased to see my gold,but more anxious to show me his good luck. He had wing damned a small cove on the side of the river to work in and when he arrived that morning,here was a big 3 to 4 pound German Brown Trout in his hole. The only way out was back up the neck of his wing dam chute and Mr. Trout weren't about to go that way. Rich took a plastic milk crate and captured that ol'boy. He spends the whole of a good mining day admiring his coming supper and telling me how to season, roll in foil and barbecue this guy. I'm gittin hungrier by the minute just listening to his description of what a wonderful meal he has in store and I don't even like fish! The final straw though, is when I go by his way that afternoon and he tells me it's pork and beans for dinner cause he felt sorry for his unsportsmanlike conduct and done released Mr. German Brown! Nice Guy Huh?

Meanwhile, Back at the Ranch or branch of the river, I start chasing a natural chute like formation in bedrock on the side of the river. This chute is smooth and recessed about 12 in. across and 3 to 4 in deep. It's a beautiful channel for delivery from underneath Big Mama downstream towards the river. To begin with,of course, this chute is not in plain view. It's covered with boulders,rocks and base material all of which required moving to expose it's existence. Close to the river's edge, it came to an abrupt halt in a bowl like basin with side walls of varying height 2 to 3 in. Now,normally this wouldn't make for any excitement because it being smooth bottomed,there's very little chance of gold staying in that bowl. However,on the river side of said bowl, there is a crack in the bottom edge running under the side wall for about 10 inches. God! Now this is exciting! Any gold coming down that chute is in that crack!! Maybe??? Well, I blew and blew with my pressurized flusher nozzle and Voila! Gold!! Flakes and pieces, but nothing big. How do I get it out, there's gotta be more in there!! Nothing I have will fit into that crack, I know there's gotta be more gold in there. What can I do??

Ahh Hahh!! A hack saw blade, in my war bag in camp just the right ticket for bending in the bottom of the bowl and sliding into that crevice. Slowly and ever so gently,I probed the hole, I could almost tell when I encountered gold (I thought). Gently I extracted two nice finger pickers, beautiful!! then Jeepers!! a nugget Big I reckon!! It looked like a monster!! And it was a nice piece of gold, not as big as I first thought with the excitement and all, but still 1 ½ penny weight ain't nothing to sneeze at. It was flat, about the size of a dime with the back edge flattened like the head of a nail. It had been hammered into that crevice and it took that hacksaw blade action to get it out. Though I tried and tried to pull out more gold, it just wouldn't happen. There had to be more under there, Jeeze, I could feel the stuff with the teeth of that blade, but no dice,it just wouldn't come. I tried wedging, and hammering and prying and cussin but no more gold was a comin. For years,thereafter, many attempts were made and failed to crack the shell of this potential glory hole. The story of course got bigger and many a miner just had to take a look and a whack at this legendary site but the ol'bed rock held on.

Some years later, I'm talking to a couple of vacationer's,Mike and Mark, whom want to dredge in the Jump Rock Crossing area. They've worked there on a number of occasions and were familiar with the Big Mama rock location and I told them the story of that hammer head nugget, even showed it to them. Later on,I showed them the bowl shaped basin and the chewed up, chipped up and much hammered crack that I just knew held a glory hole of gold. They were very much interested,however they had their own project to work on dredging, so I pretty well forgot about that crack.

Almost two weeks later, they came to me with a story. They had finished up their project on Wednesday and decided to tackle that basin crack. Now they had a railroad spike puller and had sledge hammered it into that crevice, rigged block and tackle Grip Hoist to a tree and proceeded to try and break out that bedrock covering the infamous crack. They worked at it for two days anticipating a glory hole clean up if they could just break-away that covering. On Friday morning, while under tremendous pull pressure from Mike on Grip Hoist, Mark beating with big sledge hammer,, the SLAB broke loose!! They came into the Trading Post very excited and told me that I could now sleep at night and quit worrying about that glory hole that Id abandoned at Jump Rock(It wasn't keeping me awake). Anyway Mike says they've solved the mystery and since I was good enough to share this secret with them, they wouldn't feel right keeping all the gold they found in that crack, as a matter of fact,they wouldn't feel right keeping any. He handed me a small one ounce Jack Daniel's Whiskey bottle and there inside was the gold that had eluded me and everyone else till that day!!! there was just one small piece in that bottle—a big disappointment? No not really, but the end of a good story, there wasn't even fine gold to accompany that one small lonely piece.

Chapter 9
The 1853 Silver Quarter and Preparation "H"

"Tom where in God's name does Preparation "H" mix with Gold Mining?" You know reading a title like that could conjure up a lotta things that might lead one astray from gold mining. The truth being what it is, well I reckon that the beginning of this tale is about where I should start.

Chet and Nancy Long, two good friends of ours were visiting Roaring Camp in 1986. I was shuttling back and forth to Edwards working there for a couple of weeks,then back to Camp for a week or two of gold mining. This schedule permitted me a lot of vacation at the expense, however; of reduced income. I wanted to be there for their two week stay to share their gold mining efforts, and my bosses at Edwards were very easy with my time schedule in order to keep me on.

I had just returned that early Monday morning, when Chet and Nancy had found a nice heart shaped piece of gold near Jump Rock crossing. Fran was off work that day,so we shared their pleasure and kept up our diggin down river from them. We were finding some small stuff, but nothing like their's so when I found out that they had moved across river, I moved in—a little up river from where they'd been diggin. My thinking has always been that gold can be found close to where it's been found before. I don't believe that ol' saying "gold is where you find it"and they had found this nice piece pretty close to one of my favorite area's. Surely, there's room for a little more.

Well, this time I might a been wrong, because the green stone seam that I was following upstream ran right thru a hard quartz seam coming across river at almost a right angle. This was not good news because there was no side wall room for dirt and gravel to fall down between the greenstone and the quartz. It kinda looked hopeless until I found a paper thin flake of gold (small about 1/4in) fitted tightly between the two formations. This surely didn't represent the Bonanza I was looking for, but being an optimistic gold miner, I slaved on.

About two feet up river from where I"d found that flake, the greenstone ended and dropped into a hole about 8in. wide and 10in. long. Now this was getting interesting, so I started cleaning it out. The hole dropped about a foot deep to greenstone again. It was really jammed and hard packed in the bottom, I had to break out some of the rocks. I hoped for at least some more paper thin flakes. WRONG!! The bottom was laying full of small nuggets (looked like a million—totaled 3dwt.+). Maybe not a super glory hole but man I'll take it. any day.

I didn't reason or question how it got there, I started Wiss Banging like a mad man, following that gold bearing greenstone right into the bottom of a large pine tree, man I was headin up river all the way to camp if I had to : if that gold kept a pourin in!!!. There was a huge root (about a foot in diameter) that I had to dig out underneath. The greenstone dropped down in kinda jagged broken steps for another foot under that root. The more I cleaned out the bigger it got until I had about two feet exposed of a rectangular hole that was gonna make me rich! I hit bottom and couldn't hardly wait for the final cleanout. God, I had it clean and there it was!! Staring at me an 1853 quarter that some old timer had lost as he cleaned out that hole! What a disappointment!

Man—it was time for me to take stock of what was going on here. One hole producing good and the other clean. Studying the situation, I had to take a breather and kinda determined that the ol'timers came up river cleaning the side walls of that greenstone and when they came to the hard quartz, they jumped over it to where it went under that pine tree and resumed their cleaning.

The hole that I'd found was caused by a fault in the greenstone where it jogged to the right about 8in. leaving a hole trap for gold. The ol'timers, being experienced, figured there was no reason to clean off that hard rock junction because there was no chance for open side walls in hard quartz, so they jumped over it. The staggered broken steps, where they resumed, on the bottom of that pine tree hole was where they had chiseled out the greenstone to get at the side wall deposits in the softer shiest and shale formations.

Well lessons learned by hard knocks and experience last the longest. About the time I'm ready to quit this hole and head for more promising country, here comes Norman. He say's Tom that grove of poison oak in front of you probably discouraged the ol'timers and it's probably loaded with gold. Now, I wore long sleeved shirts and gloves all thru this diggin and didn't worry too much about catchin the oak,I was more interested in catchin the gold. Wrong! I worked it about ten feet into that poison oak and there weren't a dang thing. The greenstone seam I was chasing had been chiseled and chopped out down about a foot below surface and there was zero gold. I think them ol'timers weren't put off by no poison oak.

I relocated across river and started a new area in shallow water that looked pretty good to me. It was down stream about 100 ft from where Chet and Nancy had moved to. It kinda appeared to me that this was a chute fed settling pond from up river and if anything came down that chute it had to come into that depression I was working in. So happened,of course, that Fran was off that day and was helping me—I think?—I was Whiss Banging away, when suddenly a flash of gold color appeared and guess who, grabbed the dang thing before I,The Slave Laborer, got a look see and took off upriver with said priceless gold nugget (flat round piece, about 1dwt) to show it to Nancy. (Hence the lines to a song "Pretty Girls Always Git The Gold.")

This same area produced another nice piece (about .8dwt) that I could see in a deep recess of black hard glass like rock—there was no way of reaching it with tweezers or anything resembling a crevicing tool. I tried everything in my "bag of tricks"—nothing worked,so in disgust I hit it with a hard jet stream of water from the Whiss Bangr fogger nozzle—A little bit of an angle and that stubborn nugget popped right out—Just goes to show you what scientific studied engineering and finally blind luck will do to git the gold!

Well Chet and Nancy had finished up their vacation and my two (highly enjoyable) gold mining weeks was also up, so I headed back to good ol' Rosamond home and my paying day job at Edwards AFB. I was about 150 miles into my trip when I experienced a terrible itching on both of my arms (underside) just a few inches up from my wrists. I unsnapped the sleeves and (to borrow an Elton expression) Lo and Behold, I had a breaking out of poison oak on both arms. Now don't you know this is justa bout three days from my diggin in that Poison Oak Glory Hole that Norman recommended. I had no doubt rubbed that stuff on my arms thru the slits in my shirt sleeve back of the snapped cuffs, and yes I had a big time case of poison oak.

Now everyone knows, you cain't scratch this crap or it just gets worse—a feller should wear a straight jacket when he sleeps if he wants to stay clear of scratching.

Well I managed to abstain while I was driving, but wondered what I'd do when I got home. You see Rosamond was still in the dark ages at that time, and there weren't gonna be no all night nothing open to git some kinda anti itch,so when I git home, I do a strip search of the medicine cabinet. GUESS WHAT! I find a tube of medication left over from a previous affliction of unspeakable nature (Preparation "H") Please believe me when you're afflicted, you'll try anything and the results were downright amazing.

I don't' presume to be a walking commercial for this product, but during the ensuing years since I have related this story to a number of poison oak victims and had them receive the same benefits that I did (Three day cure and instant itch relief)—I guess I've become a true advocate of Preparation "H."

So, in your future excursions, weather it be prospecting or just camping out, take along a little Oak insurance. I think I warned you at the beginning, Gold Mining and Prep "H" do mix.

Chapter 10

The Cement Mixer I Called the Prevleator

The reluctance of some individual to participate in hard work will frequently lead to innovative development of labor saving devices. This has been my forte for as long as I can remember. Hard work doesn't bother me as long as I don't have to do it.

Some good friends that were members of the Taft Prospector Club, invited Fran and I to do a little gold digging on their claims near the old Gold Mining town of Garlock California. Well, this was a desert operation which required the use of a dry washer type of gold classification unit. So we investigated the availability of one among our so called friends. Had they been real friends, they would have immediately said "stay away!" I reckon they wanted to rid themselves of this abortion, so we became the proud owners of a Mexican Dry Washer. Lord are we ever ready to sally forth on an adventure to seek our fortune near the famous town of Randsburg and the rich gold bearing area of Golar Gultch. There was nothing wrong with digging in this area. It had a long history of gold mining and we were excited to take advantage of this wonderful opportunity. Benson canyon was next to Golar Gultch and that's where the operation was to take place.

A dry washer is a dry dirt type of sluice box that blows air up through the fine muslin screen type bottom and allows this air separation to separate gold from the base material and gravel. The riffles are pointed up towards the front of the trough there by allowing capture of any heavy material like gold. A bellows is mounted under the trough to provide the necessary pulsating air. The early units were hand powered and required two people for proper operation. Clean up,of course; required the panning out of the box contents.

Fran and I are accustomed to digging a little dirt, but this sluggin away in a dust bowl leads one to question the sanity of the individuals involved. After a full morning of this so called recreational activity that supposed to be good for you, I was flat out exausted.

I drug my weary bones to the side line and watched wife Fran continue on in a very industrious fashion. The dirt was flying and I could dream we'd wind up wealthy? (Reality!)The few small flakes we panned out of our riffle box represented the sum total for our morning efforts.

Well all gold miners are optimist and, we knew that our next efforts would surely be productive (Wrong!) but that's gold mining. Our initial efforts had involved testing the whole area for color and I found a band of reddish, just a little off brown, material near a large group of creosote bushes. This material proved to be productive to a degree. I picked a couple of flakes from the pan. We attempted to work this area.

Fran had panned out about every half hour or so and in checking the contents of the panning tub, I discovered hundreds of little red clay balls that had not dissolved during the panning process. This caused me concern because I knew that some or all could contain gold. When we had finished our day's operation, I poured off the excess water and prepared to take the contents home with me. Fran questioned this because she is an excellent panner and couldn't understand me taking this material home.

About the time we were finishing, a jeep pulls up and this feller asked how we were doin. Now I gotta tell you, we are accustomed to seeing much more gold for our efforts in northern Calif. on the river, so I thought nothing of showing our few flakes. Boy these guys went ape and I knew right then I'd made a grave error, if I ever wanted to work there again. As a note,I never did because the next time I went there, I couldn't find the bunch of creosote brush because it and the road had both been dug out and ran thru a dry washer. (Goodbye to our Glory Hole)

I let my panned tailing material dry for a couple of weeks, then broke it down and ground it up. When I ran it through the sluice,it produced almost three times as much gold as what it originally panned. I was really surprised, I guess I'd expected something, but nothing like what I found. These findings triggered my curiosity and I had to start figuring a way to git the gold! My early efforts centered around getting a small electric cement mixer for $250 which my wife eyed with dubious concern and a lot of disrespect.

My thinking was to churn this red clay bearing material and attempt to get it to release gold (Wrong), then I tried mixing with rocks and even steel balls (Still Wrong) This process involved a number of trials and I thought well maybe I needed a cement mixer anyway, although I didn't have any cement work to do and wasn't about to invent any.

Well thankfully, inspiration came to my aid and I decided to try dry processing. I took all the paddles out of the drum and reconfigured the drive mechanism for a 2 1/2 sec. per revolution rate. This time period was just enough to permit the flower dust to stick to the drum walls till vertical, then of course fall to the bottom of the drum. Then I rigged a leaf blower with reduced output to blow in from the top of the front opening. The forced air blew the flower dust to the rear of the drum chamber, over to the sidewalls and out the front. Initially, I kept the pressure relatively low to make sure all the heavy particles remained inside. Later on testing allowed a much greater air flow.

Once again,my early efforts involved tumbling rocks, steel balls etc., but I soon learned that the weight alone would destroy the unit.

Once again, some one turned on a light bulb and I decided to try a different tactic. You gotta realize that this was the best thing I ever did with my 16 Pound Bowling ball. All of my gold bearing material, in the drum, had been screened thru ½ in. mesh so I figured this big surfaced bowling ball would push the clay balls thru the remaining gravel (which couldn't get out) and thereby grind them up freeing the gold. This process worked great. All the flower dust was blown off leaving only the heavies, gravel, gold and hardware. Panning out was a piece of cake.

"The Prevelator with the author winding up for a Strike!"

Now comes the down points. The bowling ball eventually sorta wears down, But I found them all over at Yard Sales for $2 (no problem)—Then is it too slow, you can only process a bucket at a time. Well to answer that, I have to say, I put the effort into building this for the benefit of two 80 yr. old friends of mine whom wanted to do a little gold mining and not kill themselves working, but still wanted to participate in some kind of gold mining activity. They loved it, played cards and ate watermelon while the ol'Prevelator tumbled blowing the dust away. Every once in awhile, they swore that the loud clink that they heard was a big nugget hitting the side wall! Pan out the concentrates at the end of the day, split the findings; and plan your next outing

Here are the instructions that go with the Prevleator—

Dig a 5 gal. bucket of ½ in screened dirt from a known gold bearing location (like in an area frequented by gold miners dry washing)

Dump this material into the drum along with the bowling ball.

Start the mixer and the blower.

Take a seat in the shade and open up a beer!

When the beer is finished, the dust has cleared, dig another bucket.

Repeat the process until you're out of beer!

NOTE: The mixer and the blower are both electric so a small gas powered generator is required (Not To Mention of course the Beer Cooler is electric also!!!)

Good Luck!

Chapter 11
Little Rich Bar

It was early July 1987, the river was still a little high, and I decided to go sluicing down river past the China Camp Falls. Turns out it was Friday and Norman's day off. I had found some good looking material up the hill aways from the falls and he came over to see how I was doin. Naturally my negative response was in keeping with my status as a genuine gold miner (unless I was trying to sell a claim).

Anyway, he was looking for a little help in getting his equipment across the river and I helped him rig up a cable on which, we hung his Wiss Bangr and ferried it across. The location of his operation looked pretty good and I was tempted to go over and test an area I liked. I stayed with the advise of an old friend (The General), however and didn't runaway from where I was digging. "Never run away from gold," he had said. Turned out, he was right, I picked up a couple of nice pieces that day.

The following week, Norman was back at his diggings working and becoming frustrated with the little gold he was getting. When you only have one day a week to work during the summer and the weather is too cold in the fall/winter to gold mine, then no gold is very upsetting, especially when you're setting right slap dab in the middle of gold country. I figured on testing this spot I liked and set up my sluice box to go to work. Now no one has a lock on gold mining luck or knowledge and I ain't sure about fishin either. Where I sat the box was dead center of the fingerling river pathway. All of these little guys are swimming up stream thru my sluice box. I decided to ignore them because the water flow was just right and I told them (the fishes) "You gotta watch for my dirt cause I ain't movin."

On this day my dumb luck was running pretty high, because the first bucket of dirt produced small 1/8 in gold flakes you could see runnin down to the first riffle. Man,I couldn't believe this kinda gold. I suspected there could be some in this cut I was workin because it had all the right kinda earmarks and around the inside bend location. Every bucket I dug was a producer and about this time I'm wishin I had my Wiss Bangr down there. Now I won't lead you to believe I was gittin rich, but boy it sure is exciting when you can actually see the gold runnin in your box. Norman came over and he couldn't believe it either, but I'd say I had about twenty good lookin flakes showin in that first riffle.

I worked a few more buckets of dirt and the only thing I got was small stuff, so I told Norman to move his operation over there. He said "no, That's your claim and I can't do that." I said, "yes it is, but I'm givin it to you," and I helped him move his stuff. He was showing gold in his box when I left to go back to camp. Not to mention, of course that he was pretty happy about it too.

Well, I moved on to greater things, I thought, and left him to clean out my discovery. Some weeks later,I went back down to that place and didn't hardly recognize the area. That Norman is a real worker and he had destroyed the whole crevice cleaning it thoroughly and getting a nice ½ oz of gold including a beautiful 4dwt nugget. I maybe shoulda stayed and worked the spot, but at that time the whole river and summer was open for me and Norman enjoyed a great time.

Two years later I'm looking for a good place to go mining, and I recall this here Little Rich Bar that I'd had a good day on. I decided to give it another go, but this time I took my genuine Wiss Bangin machine with me. You gotta understand that I had a lot more time to prospect around than Norm, so I explored all the possibilities. I broke open some seams in bedrock and moved some rock deadfalls that littered the shore line. After digging a couple of days right next to the rising bed rock wall which went up for about 30 ft. at a pretty good angle, I figured I'd probably be below water level before I hit bed rock. This didn't bode too well for my operation, because I wasn't into dredging at that time.

I was sittin there in my hole contemplating my next move because I hadn't seen but very little gold and I knew there had to be more fruitful diggings somewhere around there. Well in the process of cleaning out this hole, I'd naturally sprayed some water up on that side wall of bedrock. I'm sittin there lookin around when suddenly I see a flake of gold sliding down the bedrock wall. Naturally I grabbed that sucker before in got outta reach. Then taking a close look,I discovered a very slight depression in bedrock that this little piece was following. Here was a major revelation for me, that the surface tension of the water held onto that piece of gold. Beyond that, I figured there might be other gold up on that side wall!!

Boy was that ever right! I found gold all over the place, not big pieces 1/8 in to ¼ in flat pieces heavier than I would classify as flakes. I wound up working that wall for thirty feet up getting a nice bottle showing almost every day. I started having some difficulty,though, because the weight of my water hose was pullin the joints apart. I had to serpentine the hose winding it around bushes and rocks to keep it from bustin on me while I was working. Then I developed another problem, the water pressure from up 30 ft on that wall would blow the foot valve off when I shut the pump down, I had to go swimming a couple of times to retrieve things in order to keep working.

One morning I'm working high up on the side wall Wiss Bangin when Jackson and Bob Scruggs come down river in their little rubber boat. They had a dredging operation goin on down river and were floatin a few cans of gasoline down to their digging site. Now I was in a steep location and had been tying off my hoses and sluice box on the side wall to keep working. I also had to tie me off so I could stay there and work. At the very moment Jack and Bob appeared, I had put my holdin rope in my teeth to hold me in place. Believe me, this was tough love of gold digging. When they yelled "Good Morning' Well all I could do is sorta wave a leg or come slidin down that wall. They said later they didn't think I was being too hospitable what with not answerin and all.

The work goes on, but I soon discovered, after seeing what my broken hose runoff did, it wasn't necessary for me to use the Whiss Bang up there. A broken coupling had turned water loose high up on the wall and gold was being collected in every depression coming down to the river. All I really needed was a little water hose,a small stream carefully controlled on that bedrock wall and I was washing gold out of every little crevice and depression and crack on that bluff. This turned out to be much easier and I still recovered all the gold. The little stuff I carefully washed to a collection bowl like area and sucked it up. How much gold did I get, I have no idea, but I had a nice bottle to give Fran every day and I was havin a heck of a good time. The bluff was kinda running out on me, so I started lookin for new territory.

The river made a turn right where I was working and of course this is what led me to this area to start with. There was a bedrock ridge coming down to the water edge right at the bend and in a cut downriver of this ridge is where I'd found gold and also where Norman had Wiss Banged. He had done a real bang up job so I didn't expect to find anything he may have left. Coming right off the river was a flat pie shaped formation of bedrock about 15 feet back to a shear vertical rock wall. Facing the river and forming the outer edge of the pie wedge was also about 15 ft of bedrock shore line. On the up river side of this formation, was a column of bed rock attached at the bottom and top to mother earth. The up river side wall and back side were both open unattached and jammed full of gravel, sand and dirt all the way to the top. This was a standing wall 18 to 20 feet tall and about 4 ft thick at it's widest point. There were slim jagged spires of rock at the top end. This was an interesting formation and I was definitely getting a strong interest.

I had set my equipment up on that flat surface and started exploring the cracks and crevices that Norm may have missed. I found a few flakes and then tried to work out the bottom of that standing wall, not much to work, but in the process of scratchin out a few small flakes, I sprayed down that jammed gravel and sand behind the wall. Glory BEE!! Half a dozen gold flakes wash out around the bottom!! Man you talk about washin, I worked and flooded that wall gittin a few little pieces each time. Not fast enough, I had to git that wall outta the way.!

Next day I brought down my trusty Grip Hoist, blocks and cables,and rigged up to pull that column over. Wrong! it ain't a budging! I tried re-riggin, stompin and a cussin, but it still wouldn't fall over. I was about to pull an ol' willow tree outta the ground, but ol' Betsy was still not wiggling. I spent the best part of the day trying to figure how to bring that thing down.

Next morning, bright and early I'm sittin astraddle of that tall rock spire half way out to the point, and hammerin away with a chisel. There was a section of soft black rock tying the thing together at the very top. It was tough going and I was sittin up there beatin away when here comes Bob and Jackson again. They didn't have a clue what I was doin, couldn't hear what I yelled down to them, probably thought I'd really gone gold crazy, mining up on top of a bedrock wall. Well about two hours later,I finally cut thru that black bedrock!! The blame wall was now free standing and when I tied her off and started pullin,she came crashin down.

Around three o'clock that afternoon, I had all the large rocks and pieces cleared off. I had washed everything off blowin the small stuff back towards the wall. I sat up my Wiss Bang and got ready to git me some gold. I started washin and suckin up the debris, gravel and gold. As the bedrock cleared, gold appeared everywhere!! Man I couldn't believe it!! Gold, Gold!! All over the place, literally hundreds of small pieces, I couldn't walk without walkin on gold. I picked up some,cleaned out the box and washed the rest into a pile back in the corner of that pie shaped bedrock. It was gittin late and Fran would be worried some,so I decided to pull out for that day. Before I shut down though, I took off my shoes and washed them in the sluice box. They were tennis shoes and some flakes were caught up in the soles.

Well, there was definitely joy in Mudville that night. I didn't have no big nuggets to show, but there was a bunch of gold in that bottle. I cleaned that Little Rich Bar Pie Hole for the rest of that week and talk about good!! 4 ¾ ounces!! All pieces were smaller than my little finger nail, but Boy What a Haul!!! Thank You Mama Nature!! note: My tennis shoes were clean for the whole week!!

"But the Jeans stood on their own!"

Chapter 12
Strange Tales of Real and Some Previously Owned Gold

You know when I first got into this here gold seeking business, it struck me a lot like horse racing—You give yourself the best possible chance to win based on your handicapping ability and watch hopefully as they come in the money. (I shoulda said IF.) Gold mining ain't a whole lot different, you kinda prospect the probabilities then pan, Wiss Bang, dredge, high bank or steal as the situation dictates. Now the feller that bets on the horses tares up his losing tickets and only talks about the winners (slim though they be), "The Gold miner is reputed to be a liar standing by a hole in the ground." These following comments are some that are very often in use excusing a bad operation. "I didn't get any big nuggets, but it was just beginning to show good color and a lot of small gold during my last big Glory Hole Operation." This was just before his equipment broke down or a personal family situation came up terminating his operation. This pretty well illustrates a certain degree of dishonesty that has become closely associated with gold mining in general. Sometimes the truth is hard to come by at a Mining Camp.

The question comes up, How about Recreational Gold Mining? This new field of prospecting and gold mining that attracts people from all walks of life. The student, the housewife, computer programmers, politicians, college professors, poor folks and billionaires. Surely now this type of gold pursuit doesn't carry on the stigma of horse trader dishonesty. Well—I've gold mined with all types and descriptions and finally recognized the fact that gold does strange things to people. I'm not going to dwell on the negative conduct of some folks, but rather on the strange and sometimes comical consequences that seem to be commonplace with gold miners.

The following are a few personal experiences that my dealings with people,at the time, permitted me to observe. Sometimes you ask a question that appears very harmless and get a very informative answer that covers a tremendous amount of ground. This permits me to see the real story behind an innocent answer that following a few more leading questions tells the whole story. Sometimes I can almost give you a blow by blow description of the conversations that accompanied an incident. This analysis and interpretation gives me an insight which permits me to pass on these stories in their true form. I will tell you that the names of the participants in all of these stories I've changed in order to protect the presumed innocent.

First,let's deal with good ol'Tom, Here's a feller with a great talent for finding gold. He is a Wiss Bangr supreme and he comes upon this prospective area that he knows is just loaded with gold. It takes him four trips downriver,loaded like a blame mule, to pack all his genuine gold diggin stuff to this glory hole location. Well "Lo and Behold" to borrow a true gold miner's expression, here he gits already to fire up and his hose is a good 15ft. shy of his Glory Hole. Another trip back to camp for additional hose is outta the question, so after a bilingual attempt at self consolation—(use of expressions not contained in Webster's) He digs right there!

Yep! You guessed it, he hits a Bonanza (a couple of pennyweight in small pieces). Most of the gold was in a small bed rock fault between two dissimilar formations. Now,I know to you great seasoned sourdoughs, and other more worldly accomplishers, this don't sound like much, but for the guy that enjoys what he's doin out there—This is a Bonanza! Man if I could come up with 2dwt. that easy, you could start calling me 15 Shy! (Note: Tom got his extra hose and dug out his glory hole next day—it was dry!)

Let me tell you about good ol'Bob and Walt. These two ol'boys had been coming to camp for better than ten years and they were die hard dredgers. Dawn till dark, cold water, high water or hot weather these guys worked dredging down river about a 45 min hike over the rough shore line and big boulders. Packed in a three inch dredge with rock movin equipment and all the wet suit—air equip—etc. These guys spent one day gittin in and one day gittin out. They always had a great time,got some gold and enjoyed a wonderful vacation.

Now Walt wasn't attached,but Bob had a wonderful wife whom he thought would never come to camp because it was kinda out of her social thing. She sorta abided by the weekly hair styling, long finger nails, good wine, dining out etc. He told me that he really wanted her to come up and experience the same kinda good time that he and Walt enjoyed at camp and he was gonna try and convince her. Success?? Yes!! Too Bad!! The next year she came!!

Now Bob wanted her to enjoy gold mining, so he and Walt got the necessary buckets and diggin tools for her to work at the Caesar Diggings,in camp. Caesar"s,of course, is located up the hill behind the Trading Post and it's a pretty good chore carrying a 5 gal. Bucket of dirt down to the river to pan it out. Their first day in camp was Sunday, so his wife Flo had heard all the great Bonanza stories and was anxious to go out and find her fortune. Caesar's can be a feast or famine operation and in their case it was strictly famine. They didn't even get color and right here is where good ol'Bob and Walt made a lifetime mistake.

Bob wanting his wife to find some gold for all her efforts, salted her bucket with what I refer to as Previously Owned Gold (POG). Now he didn't salt it with a small couple of flakes, he dumps a quarter oz nugget in her bucket. He and Walt, being suspiciously helpful, very carefully assisted her in panning this bucket out. Happy!! Man that don't even come close to the smile that lady had. She had to show it to everyone in camp. We immediately had a Gold Rush to Caesar's. What a wonderful thing it was to see such a successful and happy gold mining lady. Of course Bob was very pleased that his wife was taking to gold mining so readily. He shouldn't have been!

This wonderful elation carried on thru the following day, until Bob and Walt returned to camp after a slug fest of transporting their equipment down river. Tired, thoroughly beaten and draggin in, these two miners were met by a jubilant Flo, whom just couldn't wait for them to get back to camp so they could carry her seven brim full of dirt buckets down to the river from Caesar's, and help her pan out. She had done such a wonderful job of diggin, she just knew these buckets held a fortune. Wrong!! As a note of caution, a 5 gallon bucket of water weighs about 40 pounds, dirt and rocks can weigh considerably more!

The following two weeks were anything but wonderful for Bob and Walt. I think they worked harder and panned more dirt during this time than in all their previous ten years. Flo became a great dirt digger, them finger nails were gone and the hair was pony tailed, Beer!! was good, and she was enjoying a great holiday. The gold wasn't too bad either, Bob and Walt were gittin enough gold dredging to keep Her Luck looking pretty good! I cain't print their comments concerning this whole self inflicted gold mining experience because I'm not looking for an "R" rating. (One final note—The lady has not returned to camp with them—I think Bob and Walt made sure that their future vacations were in conflict with her's.)

Movin Right Along, I bring you Al and his wife Betty. Al was a gold miner of some years experience and he was off to say that Betty didn't approve of his many away from home gold mining escapades. On this particular occasion, he had enticed her to come to camp on the expectation that she might get hooked on gold and sorta get off his back about how he spent all his spare time and cash. Well, he takes her down river about 10 minutes outta camp, (she ain't too fond of the wild down yonder, and the possible crawling wild life associated). Al sets up his sluice box in the near river shore and although he knows there's little chance of gold in this particular area, he proceeds to show Betty how it's done. She becomes quite discouraged after a couple of no show buckets and starts talking about going back to the cabin.

Al in some desperation, decides to plant a little encouragement for her. Now up to this point Al is doing OK, but he made a slight mistake when he planted this POG for Betty. He selected a wash out chute behind what we called Little Lone Rock. This chute was full of light river sand, silt and trash. Well there was a flat rock laying about midway down the chute on the surface and Al dropped four 1/4 in flakes under this rock. He then goes and tells Betty this cock and bull story about the ol'timers finding gold under flat rocks and tells her to start turning over rocks at the upper end of the chute,he'll start at the lower end.

Lo and Behold!!—Almost like the coming of a miracle—Betty finds four magnificent pieces of gold under a flat rock—Down right Amazing!!! Poor Ol'AL, for the rest of his vacation Betty has him, knowing there's nothing there, bucketing that barren sand, silt and trash to the river sluice.

Is there a moral to this story?—Yes—there is this question—Do you know what that brown discoloration is on the tail of a craw nannie?? NO? It's Mockaloomie Doo Doo!! and that's all that Al got for his helpful efforts.

Next Case Please!! Here's a two fold story that I really found fascinating because it happened with my full knowledge of what was going on and because one of the participants felt free to talk to me about it in confidence. Little did she know that this story was too good to keep secret. The man and wife team of Gail and Roger had come to camp for a long deserved two week vacation of gold mining. They had recently purchased a brand new Wiss Bangr and were anxious to try it out and attempt to find a "Book Piece of Gold."

A "Book Piece," at Roaring Camp is a nugget of 2dwt or greater. The finder's will then, upon official weigh in at the Trading Post, be officially registered in the highly regarded Trading Post Registry where their name will be displayed for all-to-see posterity and personal bragging rights. A one day find volume of ½ oz fine gold also qualifies for this recognition. This recognition is a highly sought goal of most Recreational Miners at Roaring Camp. Gail and Roger had been unsuccessful in previous years, and were frothing at the bit to make "Book" on this vacation.

Now, you must understand that Roger had been bitten hard by "Gold Fever" and Gail, understanding this, had come to camp prepared to insure that her wonderful gold mining husband had a successful experience on this vacation. First of all, she permitted him to make the studied decision as to where they would gold mine this time around. When he had finished his "Roll of the Dice" decision as to where the most likely "Glory Hole' was, she then helped him locate his new machine and started their operation.

He had located next to a large rock that had seen considerable activity over the years and unless the fifty or so miner's ahead of him had overlooked something, his chances were pretty slim. Gail, having been assured that this Glory Hole would surely be a Bonanza, then planted Three "Book Size" nuggets under the big rock. Half a day's work later, Roger was struck dumb and over joyed when he "Struck It Rich" finding two "Book Pieces" in one day. Down Right Amazing!! There's joy in Mudville tonight! The celebration lasted till the bottle was empty.

Imagine that—after all these years of hard core gold mining "Two Book Pieces."

Next day hard at it for hours and no show—not even small gold, Roger wants to quit this spot and try another on the island. No,No,No, says Gail you've got to keep going, I'm sure there's gotta be more gold under that rock. The next day still no third nugget and Gail still insisting she is sure there's more gold there. Roger hasn't seen even color for two days and finally in a raspy and angry exclamation, He says to Gail—"Look I'm the gold miner—I found the two book pieces and I say there ain't no more gold in this hole and we're moving." They moved and did find color and small gold for the remainder of their vacation, but nothing like the nuggets from the Big Rock. Gail, I'm sure cast many a wishful look at that big rock over the next few days and I know it was the last thing she saw as she left camp at the end of her vacation. I'd also say that when she comes back, that big rock is gonna be her first gold mining target.

Now normally that would be the end of this tale, but No! You've gotta hear the Rest of the Story! Two years later, Joe and Al, a couple of retired school teachers, came to camp for a little gold mining. These fellers had been to camp many times before and had found gold every trip, but never a Book Piece. They asked me about dredging around a large rock lying at the river's edge. I told them that said rock had more visitors per year than the Trading Post, but go ahead and try your luck, you never can tell about the sneaky things that gold will do. They sat up and started cleaning bed rock under that large boulder and "Lo and Behold" (again) three days later they found their first Book Piece—Now ain't that something!!

In conclusion—There are many, many tales of gold mining strikes and exploits—some are even for real—maybe??

Chapter 13
The China Camp Falls

The China Camp Fall's are located about one mile down river, maybe a little more and they are herein referred to in reference to another location for reference purposes only. Immediately upriver from the fall's is an unimposing side wall of bedrock that contains a lot of rotten limestone formation. There is a short set of shallow rapids that kinda lead into this area. Well, I had always wanted to work that area just to kinda take a look so to speak. It had no history of gold but looked like maybe a settling pond below the rapids and for that reason held some possibilities for me.

When John and Judy came to camp and expressed an interest in going down near the fall's to do some dredging /Wiss Bangin, why I naturally recommended this area to them. Now, they considered Roaring Camp a fat farm and they felt that if they worked an area that far away from their cabin, it would surely take off a few pounds just gittin back and forth. Well they set up down there and Fran and I witnessed a John disaster when he was gittn in. All of his equipment got away from him in the rapids and he had motor, sluice box, pontoons, buckets and all going down toward the fall's at an alarming rate. We were high up on the mountain when this occurred and couldn't lend a hand. By the time we arrived, he and Judy were underway working the river. Their horror story approaching disaster was much enjoyed by all the other miner's in camp and a source of a lot of friendly kidding.

There was no kidding though about the gold they were gitting, every day was a good clean up. They picked up lots of nice sized flakes, small nuggets and fine gold. Norman and I,on one occasion, drove down to see how they were doing and waded across the river, I did,he didn't. For some reason,I don't know to this day, he would go up or down river 200 yards to jump rocks across, so he didn't have to get his feet wet. Anyway, we got over there and they're telling us how great they're doin. I'm standing about ten feet in front of Judy watchin her pan out their concentrates.

She was a very good panner,so I'm sure the piece of gold I spotted under water between her feet is one she didn't lose in panning! I kinda kiddingly asked her if maybe she didn't want that piece, that I would take it. God!! She couldn't believe it, a piece that big and she hadn't seen it under water, but of course this is where they panned out and they weren't working here. I told them that if it were me I'd sure want to check this area out—They Did!! The piece that I pointed out to them was the biggest piece they'd seen so far. Norm and I shared their excitement and headed back to camp.

Three days later, John and Judy are askin, when I'm coming back down to see them, they had found more gold but nothing as large as that I'd spotted for them. Well, I wasn't working for the Camp at that time, so my time was my own, and I could go back whenever I pleased. You gotta understand though, that I didn't need no Fat Farming and it was a good haul down to the China Camp Fall's.

The afternoon did come though and Fran said if I was in the area,I might stop by John and Judy's and see if maybe I could point them in a direction where they might do a little better. Any of you familiar with gold mining, know that this is a very improbable miracle to perform. But, stop by I did and while it's late in the afternoon, they did appreciate my stopping by. The sun is behind John's back and I,m kinda standin down river of him and Judy. They're seated on a flat bedrock formation right on the river's edge resting up after the hard day's work and telling me of their finds for the day.

They've had a pretty good vacation, lost about ten pounds each and were feeling pretty good about their whole vacation. Well, I'm jawin away with them and looking at their gold in the bottle when I saw it!! There behind John on the bed rock I see gold!! This can't be true, I told him to sit still, he did, I examined the spot I was looking at, asked Judy for her tweezers and pulled out a beautiful flat nugget from a crevice in bedrock. Only the upper very thin edge had caught the sunlight and reflected the color my way. I couldn't hardly believe it, I handed it to John, He was astonished!! What a great find and him practically sitting on it!!! This was the biggest piece of gold they found during this time at camp. Happy camper's they were!!

You thought maybe the story had ended with this happy couple, WRONG!! Cause you see besides being such a wonderful feller, I'm also a gold miner. I always keep some things and some hot gold spots filed in my complex beady eyed memory. S-o-o-o on a Sunday Morning some time later, like about four weeks after John and Judy had gone home, I told Fran, I was gonna walk down to the diggings above the Fall's and do a little crevicing. She told me to be sure I returned before 1:00PM because we had to go to Sacaramento to see her Brother-in-law in the hospital. I figured I'd be back after a short prospectin run and there'd be no problem. I had no time piece, but I figured, when the sun came over head it would be about 12:00 and I'ed head back to camp.

This is what I'd told Fran. Unfortunately two things worked against me that day, the sun never came overhead to warn me and I was finding so much gold that it coulda been daylight forty and I'd probably not known the difference. I had only a small metal burnout pan with me. I didn't like trying to carry that big Roaring Camp Goldsaver Pan, although I have to admit, it's probably the best there is. I carried this little metal pan because, I'm only pannin out a half a cup full of material at any one time. Anyway when you're using a pan like I was that day you could hear the nuggets sliding around in it long before you could see them.

This is the way it was that morning and that's the way Fran found me at 1:00 PM. She was not a happy camper, She had ran down to where I was thinkling I'd been hurt,(I'd probably been better off if I was) She was thinking about doin just that when she found me!! I was probably lucky to come out of this alive!! My three penny weight of gold that I'd found that morning using just my little metal pan,didn't impress her ONE DAMN BIT!!! So ends the story of The China Camp Fall's.

Chapter 14

Nudie Wheel Diggins

On the occasion of my daughter Loreea's marriage and the fact that they were coming to Roaring Camp for their Honeymoon, I promised to take them gold mining. Loreea knew a little bit about gold mining and a whole lot about me, so I figured that she would be more than a little upset if I salted things for her and her husband. Now,it's very difficult to locate a new gold bearing area, particularly one that hasn't been cleaned a hundred times. Additionally, I'm looking for one that's relatively close to camp and to which I could pack that forty + pound Wiss Bangr of mine. A close by swimming hole and a shady hammock location would also be desirable, no doubt; but not something this gold miner was gonna look for. I went prospecting for a glory hole to satisfy a new son-in-law.

Where do I look? Where no one has looked before,of course! All right, where is that? Up the North Fork or Up the South Fork or down past the falls. The rest of the country has been scoured clean and I'm sure there's still good stuff to be found, but I'll try those area's uncommon to camp visitors. My first prospect was up river (North Fork)and although I found a little color, the access and prospects for (little work much gold) were less than very poor. I traveled down below the Falls, it really looked very promising, but much too far and too difficult for the week ahead. I went up river(Middle Fork) hoping for some success. I was looking for any indication of hardware deposits because my experience, though limited; had been such that gold always accompanied hardware and vice versa. S-o-o-o Look for hardware is what I had been doing and then prospecting the drop area.

About three hundred yards up river from the Lone Rock (a tall 40 ft or so column of stacked fractured bedrock formation) junction of the north and middle forks of the Mockaloomie, I found a place of interest. Looking up river at the walls on both sides,it was apparent that the river made a significant bend to the left. This bend was not apparent during summer water, but when that water topped five feet or so,then things were a whole lot different. I had to kinda perceive that difference and study thing accordingly. I studied the river banks closely and there under an out crop hole in bed rock looking up river was a metal automobile wheel!! Just what I'm a looking for!! It's right on the downstream inside bend of the river. This is a text book layout for gold deposits. I pulled that wheel out and jammed it in the corner of this out crop so I could easily spot it from the river.

Time for a little prospecting, On the down river side of this bedrock formation, I found a few small crevices which immediately got my attention. On all of my prospecting trips,I take my trusty bag of tricks which consists of hammer, chisel, and small gold pan. To these staple units I add tweezers (long ones), a compact mirror for shining light down into a crevice, a silver spoon(silver plated brass) because it's soft enough to bend with a hammer to fit in some tight places. Also included is a paint brush and ½ plastic coffee can lid (this bends into any crevice shape as a scoop for cleaning).Add to these a small two foot long styrene (plastic ¼ in. hose),this item permits me to blow the final dust from a crevice and uncover the gold. Here then is and was on this particular occasion my modis operin di.

I chisel and break open the crevice as much as possible, tweezer out all the big pieces, brush out the content (as much as possible) into that ½ top plastic coffee can lid, then chisel and pick any remainder. Finally I gently (very gently) blow the contents from end to end in the crevice and pick out the gold,flakes and pieces with tweezers. I failed to mention, I also carry a few gold bottles with me on all of my prospecting trips. I also take along a little fortification like water and snack. On this particular trip,I had taken a bag of Frito's. I found out the hard way that ant's love Frito's. When I took a little break from picking up gold,to grab a bite to eat,them ant's had full control of my bag. Now I'd been on a pretty long river hike that morning and I was some kinda hungry, them blamed ants had to go. I laid that bag of chips in the bright sunlight and in about ten minutes they had beat it outta that hot bag!! Boy that worked better than I had thought, them chips were good down to the last flake.

Back to the Gold! Yep! and gold it was. Every crevice on the downstream side of that bedrock produced gold (flakes and small pieces). I think I got over a penny weight just picking it outta the cracks. Man,I had to take another look at this little bend in the river. This was definitely the place I was lookin for, lots of possibilities here for a little gold. Loreea and husband should really enjoy mining with this kinda gold showing. I packed up my sack of possibles and headed back to camp.

Now,that Friday night at the Trading Post, I was telling Fran about this place and showing her the bottle of gold. Little did I realize that a mustachioed Oil Can Harry type was eavesdropping on our conversation. Later,I found out that he beat it up to my new diggin's on Saturday and hi-graded everything he could. I really don't know that he got a lot,but he implied that he did and I really resented his kinda under handed methods. Years later, as a customer at Roaring Camp; I put him into some good producing areas, and since he didn't get much in these, I figure I didn't loose too much at the Wheel Rock.

Loreea and company came to camp and I took them,with sluice box up to the Wheel diggings. Things looked good right away. We started digging on the shore line and dumping into the sluice box. Gold!! small pieces and flakes started showing and while I'm hard at it working, I wasn't paying a lot of attention to what was going on behind me. Hubby didn't know anything about sluicing or gold mining and I look around, behold!! he's laying in the stream blocking the water going into the sluice with his body and reaching into the box to pick up a piece of gold. Well, you know I could have probably enjoyed a whole week of unusual antics of this type, but I decided to take pity on him and show him how to lift up the upper end of the box to stop water flow so you could get the gold. I also let him in on the fact that the gold wasn't going anywhere, so just leave it for cleanup later when we were finishing up.

I think he may have gotten bit a little by the legendary Gold Bug, because next day,he wanted to carry that 40 pound Wiss Bangr up river to our diggin's so we could process more dirt and git more gold. He did and we did, git more gold that is. We spent the first half of that week Wiss Bangin the shore line and surrounding bed rock. I think we got a couple of penny weight, mighta been more,gave it all to Loreea anyway. We had a great three or four days there then headed back to camp with that heavy Whiss Bangr. We're gonna give the Wheel a rest at this point and digress a little bit.

During the week prior to Loreea's arrival, another ol'compadre of mine was in camp, Butch and his wife Nancy. Now Butch was a man of the cloth and a very good friend of mine. On the occasion of finding any gold to speak of, why; it was customary, at least with Fran and I, to launch our findings with a little fruit of the vine or corn field, which ever was most available. Although a man of the cloth, a little wine is good for the soul and of course Butch had a very good soul (on occasion). I had located him in a pretty good Wiss Bangin area and it wasn't long before he pulled up a couple of nice pieces of gold. This was instant "Time for Celebration." Now,I had been given a genuine high priced and high quality French Vintage bottle of the finest dark by a friend?? leaving the previous week.

Ahh Haa!! Here was a chance to break the seal and launch the gold!! We did with great expectations of imbibing a fine succulent—WRONG!!! God,that stuff was horrible!! We couldn't choke it down and the skunks avoided the area for a week where we poured it on the ground. I think Fran and Nancy shared a Root Beer with us. Let's get back to gold mining!!!

Author and Butch setting up!

Little Lone Rock is an area down river of camp, that I have already touched upon; as a location for gold activity. On the river side of this formation is a fairly large bed rock shelf covered with the usual rock/sand river debris. On some previous excursions,I had noticed a fairly large crevice(about 8in. by 2 ft by 18 in deep) at the base of Little Lone, and had kinda promised myself to explore it given the occasion. Butch had finished up his location up river, so I showed him this crevice. Right away gold started showing up and he collected a nice looking bottle from there. He fought this crevice for two days (and I'll admit it was tough going with jagged spires and broken bed rock to contend with) and given the effort involved, he; gave it up.

I came from down river and was surprised to see he'd moved out. I looked in the hole and on the bottom of the basin was a nice little clinker. It looked like it would barely fit in the mouth of a two ounce gold bottle. Naturally, I had to confirm that. It fit mine very nicely. But being a nice guy,that evening I gave it to Butch. This was after he'd assured me the hole was empty when he left. Like the ol'General said, you really gotta eye ball everything real close,or you will miss your gold!!

There are hundreds of stories out there of similar findings and similar leavings. Good ol'Gongi, a dredger of some repute, was known to come in late in the season and check out all the recent dredging holes. It was not uncommon for him to find some nice pieces overlooked by the hole diggers. On one occasion, he came in Sunday morning, and that afternoon during a wedding ceremony,in camp, here he comes walkin down the trail with a half ounce nugget he'd just taken out of a hole abandoned that morning!!!

Knowing that it was common for folk's to leave a digging without all the gold, I decided to check out good ol'Butch's leaving's when I got the chance. Well that chance came earlier than expected. When Loreea and Hubby, came back up river with the Wiss Bangr, it just naturally seemed like a good place to stop. We chopped and beat out the side walls of that hole everywhere we could find a crack, Man-n-n-n!!! was it a producer, nice little pieces of gold everywhere!! Pretty exciting!! and we didn't have to work too hard either.

Fran was there with us when Loreea was wanting to crevice some of the small pot holes in bedrock near to where we were working. I allowed her to use my trusty silver spoon that I had often found small nuggets with, while using it. I very clearly remember her digging down into a crevice not two feet away and coming out with a spoon full of compacted dirt. As she scraped the loose stuff away with her long colorful fingernail, a fat little nugget slowly came into. view Little Nugget!!! That sucker weighed in over three penny weight!! A book piece!! Man,Your luck don't get much better than this. It had to be time for another celebration. They had a wonderful vacation and headed out with a nice bunch of gold and Loreea's name forever recorded in that infamous Roaring Camp Registry.

Meanwhile,back at the Wheel, time was marchin along when I told a couple of friends, Jim and Gail of Sacramento, about the crevicing prospects there. The next day,off they go and had a great afternoon working the side walls. Jim told me that the gold was great but the heat there in direct sunlight was a killer. He told me that Gail had insisted on stripping and looking for gold in the nude. She ran around all afternoon that way? and didn't get sunburned!!. I reckon Jim didn't get much gold digging done either! Now Gail, did not know that husband Jim had told me of this out doorsy activity and later on,I related this story to Elton. He said he'd like to meet her. Said he was coming down that evening and he'd meet her then. In the meantime, I thought maybe I could stir up a little excitement, so when Gail came in with Jim, I told them how Elton kept an eye on the upper South Fork of the river,cause he could see it with his binoculars from the deck off his living room. Jim, of course was smiling, but Gail seemed some concerned. Elton, came in and when I introduced Gail and Jim, Elton gave her a very slow appraisal from head to foot and believe me she turned three shades of RED!! And that's how this area became known as the "Nudie Wheel Diggins." It's reputation has attracted a number of gold miners these past few years.

Chapter 15
"1990" It's Time to Start Dredging!

Fran worked the Trading Post and I went Wiss Bangin for about five years. She got accustomed to spending her free afternoons helping the various miners move rocks. On her days off, she had been helping me, so she knew the drill so to speak (not many women know how to rig up a boulder for moving it and darn few possess the muscle for pulling a Grip Hoist handle (type of come-a-long). One of the ladies that she helped, encouraged her to try going under water. This lady, Darlene, had the experience and the right equipment, so Fran tried this sucking air under water stuff.

Well, you gotta understand that Fran wouldn't cross the river ankle deep without I held her hand. She had tried her hand at learning to swim at a college given swimming class and sunk—literally, the instructor figured she was the worse case of rock bottom he'd ever met. The minute she lifted her arm outta water she sank. Could swim under water the pool length, but drown on top. Having been aware of her slightly less than Olympic credentials for swimming, I was completely surprised to find her going under water with Darlene. She took to dredging like a duck. That winter I had to out fit her with a dredge and all the works.

We were fortunate to get a used three inch with air equipment and the following year, we arrived in camp, ready to go win our fortune. All of my rock pulling equipment was in good shape so all we had to do was find a spot where there was lots of gold. A piece of cake! Wrong! Finding the right place to do your mining has always been and still is the hardest part of gold mining.

For many years during my Wiss Bangin efforts, I'd always taken an interest in where other folks were finding gold. On the big bar above camp I had dug a lot of holes, finding little gold while folks in the river were finding much bigger and more of it. I started to line up their find locations with small trees on the bar referenced to John''s Hole, (a tunnel in the old river channel bedrock), on the camp side and code these trees with small holes which indicated nugget sizes. I had always thought that some day this info might come in handy or maybe I could Whiss Bang in line with them on the bar.

That some day had just got here! I went to the bar and using those tree marks determined what I thought would be a good location. In the river slap dab in the middle of what we thought would be a good location was a huge boulder. We knew,of course that this boulder had been dredged around (maybe) but after examining it in the water we decided that it had never been moved. It had an egg shaped grind hole on the down river side, the lips of which were parallel to the river surface. This indicated to me that it had lain in that position for thousands of years and offered the best prospect for our project. There was about 8 or 10 inches of it above water and down about two feet there were more boulders jammed up against it. We had no idea how big it was because we had to clear it to see, we hoped it was something we could handle. All of my tree marks indicated this to be the place, so try it we must.

Three weeks is a long time to dredge and move rocks especially when you haven't seen a speck of gold. I mean a small speck!! That rock kept getting bigger until it reached seven feet across by seven feet deep and the thing was round. The only change in it's shape was that egg shaped grind hole near the top. Many times during that three weeks we looked at gold other folks were getting and often wondered if we'd maybe made a mistake. I think we convinced ourselves that since this was our first attempt, we'd stick with it till we rolled that big ol'rock because our figuring still said this was the place to be.

On Friday of the third week we had that baby cleaned all around and it was sitting up like a big pedestal on about a foot of compacted river bottom and gravel. Well Glory Halleluiah!! There was a speck of gold in our sluice box!! One little speck about half the size of a pin head! Fran and I had to launch that nugget??? (undersized tho it be) We had been a long time no see gold!! But what a time for celebrating, we could finally hope for a little color and we kinda joked about it so much that other folks thought we'd struck it big!! Don't let disappointment git you down, so we forged ahead with our plans.

Next day we hit bed rock on the downstream side of the boulder. We had cleaned out an area twelve by twelve feet to roll that baby into. Fran was running a lot of small stuff near bottom and close to that rock when I noticed a yellow rock slide off the grizzly and drop in the third riffle. I was sittin on our rock pile next to the sluice box when I saw this happen. She had only been down for a short period, so I decided to wait till she came up before I took a closer look at that suspicious rock. You kinda gotta understand at this point in her (Learning to Dredge education) that when she surfaced I'd say to her "Hey, What are you doin up here. The gold's down there." Later on in our mining activities she did throw a rock my way, it hit me in the mouth—She claimed it was an accident!!

Well about this time, Norman showed up and I pointed out this rock to him, He agreed that it looked like a flat yellow rock. It was buggin me terrible, though, because it hadn't even wiggled since coming to rest, so at the very earliest moment I shut down the dredge and guess what?? A NUGGET, the size of a nickel—My God what a find!!! Finally after all this time!! It was great!! Time to quit and enjoy—We did just that Norman carried the news to everyone in camp before we even got outta the water, everyone had to see that nugget—That rascal weighed in over 4dwt—Fran's First dredged book piece!!!

Excitement ran high in camp and we were looking forward to gittin back to work as soon as possible. The kids were using our twelve ft hole for swimming and I knew we'd have to stop that when we started trying to pull that big boulder over. I was talking to our resident Ladies Man, Jackson, on that Saturday and he kinda indicated how he'd love to get in the water again after a long number of dry years. Him being a dredger of some renown and a pretty good friend of our's, I invited him to come down the following week and he could get in the water with us if his multi patched wet suit still fit him. He was tickled to death that we'd let him do that and sure enough he showed up ready for a wettin.

We weren't quite ready to pull "Junior" yet so Fran decided to dredge in underneath the down stream side and make rollin the rock a little easier.

There was about a foot of material she had to remove and with Jack's help it made things go a little faster. She cleared bedrock with out seeing anything and most folk's would have been ready to move on, but Fran had too much Wiss Bangin experience. She scraped out every crevice and when she spotted a brown line in bedrock, she knew it wasn't natural. Diggin at this brown line opened it up into a narrow crevice and further diggin it started to show small pin points of gold. She got out her trusty tweezers and attempted to pick up one of these pinpoints and THE WHOLE DAMN CREVICE MOVED!!! The whole thing came out—It was a beautiful Half Ounce Plus Nugget—Her favorite of all time!! What a great Day!! She couldn't yell underwater but what a feeling—No way to describe it!!

She and Jack quit early that day with more and more gold

showing—many video's and photograph's tell the tale of gold—Jack couldn't believe it, he couldn't thank us enough for letting him be a part of such a great day and later Fran gave him a 4dwt. nugget from that find; to thank him, and commerate our little Bonanza. This was no doubt a great day and a great discovery for us, not to mention such a wonderful conclusion to Fran's first attempt at dredging. But, hold your horses, we weren't finished yet, I agree we had more success than most folk's dream of but "Junior" our seven foot buddy was still sittin there unmoved. It was time for us to quit pattin ourselves on-the-back and git back to work.

Work was definitely the right word to use with this baby. Fran still had quite a bit of work to do up stream before ol' Junior was ready to tumble. I spent the better part of two days just riggin up the rascal. But finally that afternoon came when all was ready for the big pull and we both had our Wheaties for breakfast (I Think). That sucker wouldn't budge, we busted a shear pin or two trying to get movement—no dice. All six blocks were in place and that ol' grip hoist line was a singing, but still—no dice. We quit for the day, Left our lines in place, Took the handle out of the puller body with full tension on that rock. This was not the safest thing to do, but it was close to gittin dark and we warned all in camp to stay clear, especially the kids.

Next morning bright and clear, I looked upstream and guess what?? My lines are drooping into the water. Glory BEEE!!! that big ol' bad boy has turned loose. We pumped the big tumbler with one hand. The river and all night tension had done the job. She rolled like a big ball over on her side into that big pool we had built and still lies there to this day, just barely underwater during the summer months. It was now time to see what treasures were waiting for us.

There was gold in the dredge and Fran would occasionally bring a nice piece to the surface. Many times she made book (pieces over 2dwt) during this operation and her bringing up a book piece became so common place that I got scared, I was treating this too nonchalant. There had been too many dry (no book) years for me to be so unconcerned about gold.

We were using a solar oven to cook our meals at this time and I had to make occasional trips back to camp to rotate our oven towards the sun.

Lots of folks in camp did it for us but you couldn't always depend on that, so whenever possible, I'd run back and check on dinner. You can't believe how well those things work, but that's another story. I come rushin back up river to our diggings, I didn't like leavin her alone but she wasn't movin any rocks and there were other folks close by. She comes up outta that water with a jaw breaker smile and ask's me "Is there any one else around" I say "Nope" and then she opens her hand and there lies the prettiest nugget I've ever seen. It resembles a figure holding a child "The Madonna" in my mind, it is absolutely Beautiful!! This Baby weighed in over ¾ of an ounce!

She had found it in a dead end cavity filled with sand, directly under where Big Mama (the rock) had lain for thousands of years. Happy don't come close to our feelings, what a fantastic three weeks of dredging. There couldn't possibly be another to match it, but I'll tell you we were gonna keep trying anyway. Our stay at camp wasn't up yet, so we hunted around for some more gold. Didn't have much luck but then we weren't too unhappy about anything at this point, life was good, we'd found lots of gold and we were there to enjoy. We Did!!

Now comes a little problem, we'd worked our way down stream about 10 ft. past big Mama when we found a cemented crevice in bed rock. It was down seven feet and we had to beat the material out with a sledge hammer. Here was more Gold!! Bunches of it!! We'd go about a foot beatin away and there lay another penny weight or so. All mixed sizes and once in a while another book piece. To say we were happy camper's is putting it mild. But our luck ran out—not the gold—just our luck. Fran had injured her arm at work the previous winter and tendonitis was really gittin to her, she couldn't swing that ol'hammer anymore. She came up in tears a few times because it hurt too much to go on. I couldn't work down at seven feet because of an ear pressure problem. I'd had an operation as a kid and it messed up my right ear. Just couldn't take the pain below five feet, so I stood in the water, weight belt on with my air and beat the bottom out with our long breaker bar. This worked OK for a couple of days, then my arms start givin out and I've gotta play music a couple times each week. We had to quit the hole and relocate to shallow water or quit mining.

Our daughter Loi, was in camp with us at that time, but she was just about two months shy of giving birth to grandson Lucas. It hadn't stopped her from dredging, but we had a heck of a time putting a weight belt on her heavy enough to keep her under water. Not to mention of course that my face mask didn't have enough room for her nose. Wonder where she got that from?? Anyway,she was OK on top or in shallow but couldn't help much with that beatin bottom stuff. We moved down stream.

Fran and I decided we'd try the upper end of the island straight out from the first cabin upriver of camp, "Bonanza." This was a pretty wide swat and there was no telling where to start. We decided we'd start on the far side and cut a four foot channel across on bedrock. There didn't appear to be too many big rocks and we felt this was the best way to find a deposit or drop zone. It was only three feet deep and Fran went across pretty quick, about two days dredging. I checked the sluice box about every half hour or so, a quick look in the first riffle for color. Early in this effort,I noticed a little hardware and river garnets in her box. There was no other sign the remainder of that week and when we finished, we decided to go back to that garnet area.

There were two little trees straight down stream from us and that looked like the way for us to go. We sat up to do just that when our Aussie friend Sally came into camp with her family. Now these were very,very special people to Fran and I and they only had a weeks vacation. We offered to let them use our equipment and dredge out the hole we'd started, Tim and Sam (Sally's boy's) loved it. They were turning blue in that co-o-o-ld water but hung in there to learn how to dredge. We had them keep what they found, it wasn't much, but I think there was a small nugget and some flakes.

Well our time was starting to get short and we kinda had to get serious or start packing up. I'd go down at six in the morning and roll rocks into the river off the island in the direction we wanted to go. It then took us almost two full days to pull them out and hand pick all the small ones. On the third day, Fran would go down and clean the bottom. We started finding gold immediately, nice pieces, some book and a lot of bottle rattlers. One really nice 8dwt nugget!! Boy they don't get much better and here we had found so much already.

Then one afternoon,we were cleaning out the hole and here she comes up with a beautiful flat sculptured nugget that is a dead ringer for that Alf Creature on TV. It weigh's over 4dwt and is already open for a necklace chain (not that hanging it that way is a good idea due to the gold softness and wearing through hanging on a chain). Fran loved this nugget, but when Elton"s wife Delores saw it and expressed the wish she could find one like it, because she didn't care for those bigger globs for a necklace, Fran gave it to her. I think that Elton was shocked, He shouldn't have been, because Fran and I both thought the world of Delores.

Our stay at camp was coming to a close and Wayne Smith and family had come to camp. He was asking about where to dig,I told him about the hole Fran and I had abandoned because of injuries. He decided to dig elsewhere, much to the disapproval of his son Todd. Three days later,He asked me where the hole was, I showed Todd. By the coming Saturday, they had taken out over an Ounce of Gold!!

I was happy for them and I can still see the smile on Todd"s face. We hadn't done so bad ourselves and we still had a hole to clean out before we left camp.

This was our last Friday, we were packing up Saturday to store our equipment in the barn,so I had to be out of the river Friday night. We had scheduled our work so that Friday afternoon would be Fran's last clean out. Then our friend Jack M. of Los Angeles and his two young boy's were gonna move into our hole because it was such a good producer. I'ed helped Jack get his dredge into position so all he had to do was float it in. A lot of other dredger's in camp were just as ready, but of course,he had the first shot at it. I drug a couple of pretty good sized boulders out that afternoon and Fran was cleaning up the debris before checking out bedrock. Now lots of folk's think that the ol'dredge will just suck up everything in sight and I've had people tell me how that twelve inch dredge will take a man's arm outta socket. Well all of this may be true, but; it's also true that it won't suck up a one ounce nugget laying on bedrock, so unless you can see the bedrock and physically see that it is clean,you may leave your better gold laying right there. A very learned professor once sent me a report which stated the requirements necessary to generate enough suction to pick up gold and unless there was some way to get large nuggets into your suction hose and have the gravel and sand push it to your sluice box, then feller you ain't gonna git it. Case in point!!

I told you we rolled rock and debris for two days, then cleaned up the bedrock we had cleared. This was our method of recovering gold. Fran always spent about three days moving boulders and over burdan, then went down and picked up gold and cleaned the bedrock for about one hour. During the first fifteen minutes of this hour is when she got most of her gold. This last day, we had planned to do just that, Fran was ready to go down for clean up when someone up river dredged into a purple colored decayed rock. The whole blame river became a milky white and since we were diggin into a cove on the point of that island, our hole was stagnant and wouldn't clear out. She pumped and ran our suction hose continually across a blind bottom of bedrock about six feet by six feet but couldn't clear the hole. Finally she had to go down and feel across the bedrock bottom to retrieve all of her tools. There was very little gold in our box, so we packed her in and pulled Jack's dredge into place.

The next morning when we are packing up, after taking all our stuff to the barn; here come's good ol' Jack with tidings of great Joy. He is showing us a bottle of gold with two book pieces, the total of which weigh's over ½ ounce. He had only picked up what Fran couldn't see and he never moved a rock!! He said later and still does, it was his best year, ever, at Roaring Camp and he'd picked most of the pieces up with his fingers! We were happy for him and his two boys. They were some kind of excited!!

This was the climax to a fabulous gold dredging season at Roaring Camp and a very gratifying season for Fran and I. Some of our very close friends had enjoyed and shared in our success, and we had packed away a lot of great memories. What can you say, we could have had all the gold!! Then What?? The next year, I told Elton, Fran and I are not diggin the island anymore, we're gonna let someone else enjoy.

They did and we did!

Chapter 16

All the Luck

Incidents are not an unusual thing for Roaring Camp, but this particular one was a very narrow escape for a couple of regulars. Fortunately, all the Luck at gold camp is not limited to finding gold! The year was 1998 and the river ran high right up until the end of July. We had six Common Operations that year and the July event ended around the middle of the month. A number of customers had scheduled consecutive weeks in order to participate in the Common and then engage in their own private operations. This was the case with Rich and Mark, two experienced gold dredgers, whom were regulars that normally spent multiple weeks at Camp gold mining.

High water is a frustration for gold miners because they can't get in the swift water with any kind of safety consideration. At this time the river was running five feet above normal in the swimming hole and of course that didn't look so bad to these miners. Above camp though was a totally different matter, the water was much higher in the narrows and very, very dangerous. They had asked me about crossing the river to get to an area that they thought was loaded with gold. I told them definitely NO, it was way too risky. They chose to ignore my warning, the area they wished to work was just across on an island above camp and looked easy (if you could get there).

They went down river to the swimming hole and swam across wearing their wet suits. They entered the water on the upstream end and were dangerously close to the rapids when they finally made it across and onto the bank on the far side. Mark was having the most difficulty because his wet suit wouldn't zip up his back (he'd picked up a few pounds since wearing it last) They had,however made it across and now to get upstream and try to get some equipment over to prospect their Glory Hole.

An associate on the Camp side used a fishing line cast over attached to a small rope which they tied off on the island. This sounds a lot easier than it was because Mark and Rich spent a good half hour getting to the island. They had to earn every inch holding on to trees and fighting a current they now realized was a lot more than they bargained for. Finally getting the rope and tying it off, they decided the current was way too swift and strong to attempt to pull a dredge over to the island (This was smart thinking—for the first time that day)—Had they tried such a thing, the river would have owned a nice dredge. Well now that we're tired,cold and exhausted, how do we get back across river? The swimming hole would have been the best choice, so they naturally decided to try upriver.

This was not a bad choice for Rich, he was a good swimmer and he went up about 300 yards above camp. He had waited, forever it seemed, for Mark to get in and finally Mark headed further upstream looking for a safer crossing. Norman was on the near shore and followed Mark from across the river. Rich, in the meantime, jumped in and swam hard against the current fighting all the way. He made it across to camp side shore up stream about 30 yards above first rapids.

I had been up camp earlier checking on their progress, when I'd learned they'd ignored my warning and were across river. This was a good three hours they'd been slugging it out with the river. I walked by Rich's cabin, he was sitting there on a rock looking plenty exhausted. I walked on up river and spoke to Jim (another camper gold miner). I asked about Mark and was told that he was still up river trying to get across. I headed back into camp when I saw Rick's wife Gloria. I asked about Rick, I was gonna try to find out where he had left Mark. Rick had headed for a hot shower.

I looked up river where Gloria said Rich had jumped in and at this point everything went to hell in a hand basket!!! There in the middle of the river was a head coming downstream at about 30 mph!! It was Mark!! Making no effort to swim, just floating in the swift current. I ran up to John's Hole (the diggings up river of camp) and tried to break a limb off a pine tree buried in the bank (no luck)!! Norman was standing on a high point above me yelling at Mark to swim! Swim!

Mark wasn't moving! Just blasting down that river headed for sure disaster in the rapids straight ahead! I don't even remember considering anything, I ran down the bank, out on to a projecting boulder and jumped as far out as I could to try and grab him. I luckily caught his right fore arm (the wetsuit) with my left hand. As I went under water when my feet hit bottom,I jerked him towards me as hard as I could. The current and his weight (220+) about took my arm outta socket. This effort had to be blind luck or somebody was helping me out, because he came straight at me, grabbed me when I surfaced and said "I gotcha Tom,I gotcha." Well he gotched me okay, he was wound all around me. The river spun us against the bank, I grabbed a pine tree limb and hung on, the current hit us, the damn limb broke! Here we go again! Another limb shows, I grab again. This time one of Marks legs goes over my head and locks me against the bank, the current is pullin hard, my head goes under water, and I,m struggling to get to the surface and breathe, but Mark has me pinned under water. As this limb starts to break, someone grabs us from the bank! Jim and Norman are there to help us outta the water. We slowly unwind and climb up the bank.

"Thanks Tom" and Mark gives me a big hug. Man I'm beat to death, hurtin all over from the rock bruises to my legs and I just wanted to get to the Trading Post and a hot shower. Fran was scared when she saw me, knew that something terrible had happened, but was willing to wait for the whole story. I showered and then the reaction set in. I couldn't help myself—I couldn't stop crying—I guess a counselor would have a good explanation for this but right then the reaction had hold of me and every time I thought of Mark almost gittin killed, I broke up again. I was finally able to get the story out for Fran and then I settled down. Mark came into the Trading Post a little later to see if I was OK. I told him yes,but I sure need a drink(even though I don't like whiskey) and I insisted that Mark join me although he never drank. He said on this once in a lifetime occasion (and we surely hope it's only once) that he would share a drink with me. I think we had a double slug of high octane bourbon and celebrated the many good times to come.

The next day Gloria, Rich's wife, whom had witnessed the whole thing said, "Tom, you're gonna have to leave this kinda stuff for someone else, maybe you don't know it but you're gittin too old to be jumpin in the river like that." Thanks Glo!!

Chapter 17
Rattlesnake!!

You know anytime I get near a rattlesnake,he's immediately on my endangered species list. It was therefore quite a surprise to me in 1985, when I heard the Environmental Protection Agency of Amador County had placed the Brown Backed

Valley Rattler of Amador County on the endangered species list. Why, I just couldn't believe it! I know that in 85 the EPA Bureau in Washington was enjoying the highest degree of recognition and self assumed importance that they ever attained, but this was no reason to do such a thing as practically inviting these dangerous critters in your home. I figured somebody has sure got their wires crossed somewhere or they're smoking the wrong stuff. Why, every rattler I killed last year was a Brownback and we can't let them exist where children and dogs are running about, much less preoccupied gold miners.

When I heard that Bob Little, our local Game Warden was coming down to Roaring Camp; I immediately hustled into camp to ask him about this EPA business. He had come in with one of his Deputies or Eagle Scouts or whatever you call those folks. Anyway, he wasn't there but had gone down river to measure the water temperature at the junction where the North and South forks of the Mockaloomie come together.

I went down and encountered his deputy near a place called Lone Rock. He had his instruments in the water and was evidently recording the values. I accosted him with this question about what the bureaucrats in the EPA had done to our rattlesnake protection—they weren't protecting us and (sarcastically) do we call in the State Militia when we see a rattlesnake. His answer was, "No Tom, but we the Fish and Game Rangers are obligated by oath of office; to cite you if you're seen killing a Brownback that's under two feet in length and that's a $500 dollar fine. It's the Law!" I told him, "You folks are crazy!! it's the craziest thing I've ever heard of and I'm still gonna kill them,but; I guess now I have to watch out for Rangers too!" I said "Where is that Bob Little, I'll take this up with him" He pointed down river and I took off in that direction not paying real close attention and mulling over this dumb law while I made my way.

Now back in 1983 we had a real cloud burst gully washer come down that wild uncontrolled South Fork. The high water curled around a bend down just above the Falls and washed all the sand away from a grove of young pine trees growing there. This wash out left all the upper roots exposed and I always cautioned folks to be careful walking in this area because the rattlers like to curl up between those roots out of the direct sunlight and lay there during the heat of the day.

I was making my way thru this grove when I saw Bob Little on the shore line about 50 yards down river. I yelled at him and waved to get his attention—about that instant with one foot in the air a Four Alarm Rattlesnake goes off practically under my foot—I am Petrified—Frozen in motion—anyone that ever came that close knows exactly what it's like!! I am frozen in motion and I'm facing the biggest rattler I ever ran into. I reached down,grabbed up a root, and started beating that sucker. I flailed and flailed away at him. I noticed Bob running my way for all he's worth. I kept beating and beating and finally stomped him to death. I am soaking wet, shaking like a leaf, catching my breath, standing there holding onto that root when Bob finally gets to me. He stops, starts shaking his head and begins taking his Citation pad out of his pocket and says, "Dammit Tom,I sure hate to do this!" I yelled at him "What the hell are you talking about? Bob!! Look at him!! He's over five feet long and big around as my arm!!—Then Bob says, in a slow methodical way; "I'm not citing you for "Him"! Tom, I'm citing you for that little one you grabbed up like a root and beat Him to death with!!!!!

Well—of course—it's all a lie—but here's the "Rest of the Story"!

In 1995 at the Liar's Contest in Death Valley, I won First Place with this story and they wanted to award me with this beautiful Silver Plaque saying as how I was the Champion Liar of 1995. Well I stopped the awards when I told them I couldn't accept the trophy. The M.C. said, "Why Tom? that was a great story and one of the best we've ever heard!!" I told them, "I can't accept because it says here this is an Amateur Contest and I'm a Professional Liar!!"

Well, I never said that all gold miner's tell the truth, but I will tell you when it's not. The following stories are fairly authentic, especially those in which I was a participant.

Bob Scruggs and Luck's Cabin

About the mid-sixties, a Jeep trail had been cut into the side of the mountain and it was now possible to bring customers into Roaring Camp without the former necessity for coming down on horseback. Bob, a good friend of mine; and a noted gold miner of extensive experience was working for the Mining company and had been shanghaied into escorting a small four man group of visiting would be gold miners and teach them a little about mining techniques. He had been directed to sleep out in the ol'Luck's log cabin. Now Luck's had been built around 1850 and I believe to this day it coverers a mine shaft going back into the mountainside which accounts for the weird noises still heard in the night. Additionally the stories associated with Luck's gave caution to anyone wanting to sleep there. Later years had seen it rebuilt and modified frequently and at this time it wasn't in the best of condition. The dug out basement was full of old mining equipment and was falling in from the side walls what with the burrowing of ground squirrels and pack rats. This, however; was the only place suitable at that time which would accommodate four would be gold miners, a guide, thirty odd bats, a host of ground squirrels and untold numbers of pack rats. The live stock, was an added attraction for city slickers. After all, "Nothing is too good or too unsuitable for a gold miner." If you were looking for original authentic gold miner adventure; Boy, you came to the right place!

The mining work had been miserable that day and although the evening cook out steak was great, Bob had to do most of the mining and cooking and was one tired pilgrim that wanted to head for a bunk. In those days, you cleaned the leavings (pack rat homestead) out of the bunk box, filled it with new clean leaves or grass, threw your blankets or sleeping bag on top and crashed in for the night. Bob was kinda oblivious to the complaints of his party and in those days his temperament wouldn't stand too much gaff without a physical explosion. His group recognized this and kinda softly went about grumbling and crashing in as it got dark.

"Sweet James" as Bob referred to him; was one of the visiting group that had probably slept on clean linens most of his life. As soon as the lantern was blown out, the pack rats came out in numbers looking for anything to eat. The new smells of food and equipment required an immediate investigation. The bats were out swooping thru the cabin and the noises were petrifying Sweet James. He loudly complained, "What's that noise, What's that noise?!!" "I can't sleep with all this noise!" Bob told him to shut up,that the scratching and shuffling across the floor was only pack rats and if he stayed in his bunk, they wouldn't bother him.

Sweet James sat up quickly with "OH My God!,"but quieted down for a two whole minutes and he was back at it again "God! I can't sleep with these rats," he complained. "Shut up," once again

Bob told him."I told you they won't bother you if you just keep your arms and legs in your bunk, now Go to sleep!" Things were quiet for almost five minutes, then once again Sweet James let go with a yell "I can't sleep with these Damn Rats and bats running around in here!!" This time Bob had enough, he got out of his bunk, went over to the big trap door that covered the dug out cellar, lifted the door high and let it slam down on the floor. Three or more rattlesnakes started buzzing in the cellar and all the pack rats bailed out. "Sweet James had no more to say."

Bob told me he then, had a good night's sleep but; he didn't know how the other's fared or if they even stayed in the cabin that night.

Now, I won't lead you to believe that Roaring Camp proper is over run with rattlesnakes, but it is wild and wooly untamed back country still very much like it was in the 1800's. I have worked and vacationed here for twenty six odd years and during that time I've enjoyed a lot of experiences, some of which involved the wild outdoors just like it used to be. Civilization has displaced a lot of our wild life but, there are still virgin areas (not near enough) where one can step back in time even for a little while. The following stories illustrate some of the rattlesnake humor I've been a party to.

A group of my Rosamond neighbors came to Camp for a gold mining vacation and being the neighborly thing to do, I led them on a trip down river. They had located their mining effort adjacent to a rock that looked like a large thumb. So this group became known (to this day) as the Big Thumb Mining Company.

They had heard of the Wiss Bangr and were wanting to try their luck with one. I distributed the parts and pieces among the group for carrying to their mining area. Some carried buckets and equipment etc. and I carried a large heavy pry bar for moving boulders. Les (a good neighbor and friend?), I had him, follow close behind me with a long handled shovel. I cautioned all that we could possibly, encounter a rattlesnake in our hike; and I wanted Les close by(with the shovel), for that reason.

We struck out from camp and forded the river down stream, at a sorta rock hopping crossing. Now I had gone this way hundreds of times without incident, but on this occasion; when we stepped on shore,a big ol'rattler took off up the bank: crawling under a large flat rock next to a large oak tree. No one of this group wanted to stay in that area. They wanted to vacate muy pronto, but I told them we had to get Mr. Rattler out of our crossing site or the next party might not be so lucky to see him,and could get bitten. I asked one of the ladies to get me a bucket of water and asked Les to stand by with the shovel. I told them that I would pour the water under the side of the rock and Mr. Rattler would come out headed for high ground(because he would react as if the river was rising) and at that time Les could deal with him. I poured the water, out came Rattler, but where's Les? Standing on a rock twenty feet away, out in the river! I had to use that heavy bar to take care of Mr. Rattler. I then asked Les,"What was your problem? Why did you hang back there, I needed you to whack Mr. Rattler for me!!' He said "Tom you're a good friend and I didn't want to let you down, but; this five foot shovel is exactly five feet short of how close I'm ever gonna git to a Rattlesnake!!"

Well, I'll agree that Rattlesnakes do make some folks nervous and some kinda over react a little bit. Friends Dave and Marleen abandoned their diggings when they found a Rattler was more at home in their "Glory Hole" than they were. They even left their tools there just in case he wanted to do a little mining on his own!

Then there was Tony and his girlfriend, that asked me to set them up with a Wiss Bang up river. They wanted to be out of camp(possibly for some other reason) and I took them to a place where they were secluded and could maybe find a little gold. It was a great gold mining area and we found a couple of flakes after setup and running for half a hour or so. They were both ecstatic about the location and their good luck. Tony dropped a flat rock off the sidewall and guess What? A small rattlesnake zips out of there. Well

I took the spray nozzle and flushed him out of his hide out and dispatched the critter.

Tony thanked me and I headed back to camp.

That evening,I asked of his day's luck and he told me that the gold was great but, they had moved down river because they couldn't look for gold as it was just impossible to watch out for rattle snakes at the same time. I asked where he had relocated to and he described a large ditch with bathtub like rock basins on the river's bank a ways downstream of camp. I didn't have the heart to tell him that he had relocated to the area known for over twenty five years, as Rattlesnake Gulch!

Then there was Joe and the Plummer. Here are two ol'boy's well known for their gold mining exploits and long time visitors to Roaring Camp. They were dredging up river and had come to camp to enlist my aid in showing them where to locate for the best possible chance at finding a little gold as a lot of the river in that area had been covered pretty good. On this particular excursion, "Marshal' Neil (famous Rattlesnake eater) was with us along with others. As we passed through the up river gate, I advised all to keep their eyes open for possible rattlers; because this particular area was traveled heavily by all kinds of snakes trying to get to the water.

The line up was single file and no sooner had we cleared the gate area than boom! I bumped into the back of Neil whom had bumped into and all in front of him. Snake! Was the word passed back. I moved to thc front and here was Joe noted Las Vegas poker buff and bouncer,(250 plus pounds of muscle) with a hundred pound rock hoisted over his head to deliver a goodbye blow to a Rattlesnake laying across the pathway. He says, "Tom,do you think this will kill him?" About the time I was gonna say yes, he launched that missile and severely disabled Mr. Rattler. Before he could grab another rock,I moved in with my walking stick and pinned the head for Neil to conduct his butchering job. Now,Joe was some put out because

I had stopped his attack on the snake but, I told him that I didn't want him grabbing another rock because the hundred pounders he was likely to throw would completely tare out the narrow little trail we all relied on to get up river. Besides, Neil wanted more than Rattlesnake hamburger for dinner!

There's another rattlesnake tale that I hesitate to tell because, it reveals a part that I played that some parties are unaware of. Fran and I were coming into Camp at just about dark one evening with frozen food we had picked up at the local grocery. This meant we were in a hurry to get down the three and a half miles before our frozens thawed out. We were not speeding at 12 mph but were not dallying around sight seeing. Swinging around a curve on this narrow and steep dirt road, I encountered a snake stretched across the road ahead of me. I hadn't seen it in time to avoid and ran over it. It was fairly long and large,so I took it for a gopher snake. I told Fran, that I hated to kill a gopher snake but it was too late and that the snake would probably crawl off the road and die. We went on into camp and forgot the incident.

Later on that evening, Shirley, the Boss Lady of the Trading Post and totally petrified of any kind of snake,came into camp. She had a harrowing tale to tell. She came swinging around a curve in the road on her way down and ran over a huge Rattlesnake coiled up in the road. She stopped her truck thinking she had straddled the snake but, knew that she should get rid of it anyway. She reached in the glove compartment for the flashlight that they always carried; "Damn you Norman!" (her husband) she exclaimed! The flashlight was missing! She backed the truck up and there in the headlights was this big snake still very much alive but, unable to crawl off the roadway; she thought that she must have run over it! Shirley climbed in back of the pick up to get the shovel they always carried there just for this purpose. "Damn you Norman!"she once again exclaimed! The shovel was missing! She had to get this snake off the road before people came in on a day trip the next day. Bosses Orders and it was known that she would be charged with this error, if left there; so Mr. Rattler had to be removed.

With the aid of a few more Damn you Norman's,she drove the truck over the snake four times just to make sure that Mr. Flat Rattler was good and dead. Picking him up by the tail,she threw him in the bed of the truck and upon arriving at the Trading Post, she informed "Mr.Damn you Norman" that he had a snake to remove!

As for me, I sat there and listened to her story, wide eyed; with the proper amount of astonishment and admiration for her bravery. I had no wish for her to trade part of her anger to "ol'Damn you Tom," for leaving that crippled Rattlesnake for her to deal with.!!

There are many incidents that occur at a gold mining camp that generate a sideways look at the humor involved. You've all enjoyed the antics of the Three Stooges Comedies and this story kinda fits that category. This particular incident involved our snake eater Neil and another vaunted gold miner Mr.(hard of hearing) Joe Dyke. They were in the process of heading downriver to a favorite mining site and Neil called upon Joe to lead the way and listen for rattle snakes. Joe turned around and yelled back "What did you say?" Neil once again, cupping his hands around his mouth, yelled "You lead the way and listen for rattle snakes. We're right in the middle of their country!" Once again,Joe cupping his ear said, "What did you say?" Neil, walking up to him very loudly, said, "You go first and listen for Rattle snakes!" "Oh, OK," Joe says, "I'll keep a close eye out for those critters!" Consequently, Joe walked right by a big ol' rattler without a glimmer of knowledge that he was there. Neil, coming close behind; almost stepped on the critter. "Damn you Joe, he yells, didn't you hear this Rattlesnake?" Joe came back to where Neil was and said, "No and I still don't hear no Rattle snake!" Neal says, "Here bend over real close and you can hear him real good!" Joe, of course had a little more sense than sticking his hard of hearing head in the den of a Rattlesnake. They were standing within less than two feet of this Big Rattlesnake that fortunately decided to crawl back under his rock before Neil's appetite over came him.

Neil came all the way through camp and up river to tell me about this large Rattlesnake and to warn all in camp' of his presence down river. I told him to grab a couple of buckets of water and follow me because I didn't want to leave such a danger in the path of our vacationers. I picked up a long handled shovel; along the way.

We came upon the rock where Mr. Rattler had retired, and I poured one of the buckets under the rock. No Response! I relocated to the up river side and poured in the other bucket. Boom! out came big Daddy, madder than a wet hen and ready to face Marshal Snake-eater Neil and any other snake bait handy! I slammed his head against a rock and held him penned until Snake Eater Neil could direct his spirit to the Happy Hunting Grounds. Forty one inches, for our country; a large Rattlesnake and sure to please the palate of Snake Eater Neil!

Well, Neil; enjoyed a great breakfast of critter and grits, then proceeded to prepare this large snake skin for a possible hat band. Joe Dyke had a different view of the use to which this skin should be put. Considering Neil's recommendation for sticking his head in the Rattler's den to hear the rattle. He said, "If the skin makes too large a hat band for your big head, Neil: Maybe we could use it for a Hangman's Noose for around your nasty neck!" Hey Neil, you are truly adored by your fellow workers.

This common knowledge of Neil's affinity for Rattlesnake cuisine has led other campers to post signs on the trails when they see a Rattler. Most recently a paper plate was found hanging on a tree upriver, Warning of a Rattlesnake with an arrow pointing the way. Joe Dyke says "They should have left a knife and fork to go with the plate. I hate to see Neil eating with his fingers!!"

Chapter 18
Fountain of Youth

This little creek carries a name that holds lots of memories for me and it would take considerable time to tell'em all. What's the Fountain of Youth?? Well it's a spring that runs outta the mountain about half way down the road into camp. I always tell folks when we stop there, that you take one drink from this spring and you may not get a day older!!!. Did I say that wrong?? OH, Yes! and please don't drink from the tub under the feed pipe, all of our animals drink outta that and we don't want "Them" to catch anything!!—Well I reckon this is supposed to be some kinda gold story, so I better git on with it!

Roaring Camp in 1996 suffered a 500 year flood which moved things around considerable. Since I spent a lot of my time running up and down river while working with customers, I had the opportunity of seeing all the changes made by high water. Across river from the Fountain of Youth Creek outlet, I saw a change in the river surface. Examination proved it to be a wash out of the original shore line and taking a very close look, I got quite excited when I determined that a small ridge of bedrock was showin just above ground on the new bank of the river.

This would be a perfect place to do a little digging because I'd found some pretty good gold very close to this place some years back. The only trouble with this find was, it was too late in the year to start a project on the shore line that would require bringing in a water source to do dredging. Wiss Bangin was outta the question also because of the volume of material it required moving, so I decided to file this info away and hope for a better shot the next year. Fran and I were pretty well engaged working for camp over the next few years, so we didn't get back there for awhile. I didn't see any reason to put anyone else in that spot because most folks didn't have a long enough stay to get the job done and we were too occupied running camp to work that far up river.

The tail end of 97 came early for Fran and I, we'd done a little dredging below camp, mainly just to keep a hand in and we really hadn't seen much gold in our operations. We'd dredged into a dead end cove on the bar below the swimming hole and aside from Fran hitting me in the mouth with a rock, why things were pretty quiet. Aussie Sally and family came to camp during this period and wanted to work across river, Wiss Bangin on that same bar. We may have encouraged that because they had no rock pullin equipment and we figured we might could help'em out a little. I think though, the high light of this operation was the two burlap sacks I filled with bubble wrap to make a float for their motor, to get it across. It carried a sign appropriately stating the name of this river craft as the

"USS Mockaloomie" In a later year this name carried a very important significance for Sally as she flew the Stars and Stripes on her little ship to celebrate the award of her citizenship.

It having been a bad gold year for us, I decided to try something I'd always wanted to do, that was to clean the white marble slab on the island below camp. It was October and we had to clean up and close out camp before we pulled out. Early mornings I worked and at noon or so, I'd go down river to Wiss Bang the white marble. This slab, as I refer to it was a huge formation of marbled limestone that surfaced in the river channel below our swimming hole. Most of it was covered with boulders, gravel and sand material. In previous years,I'd tried to clean it in spots. Most of the time with limited success and without the aid of my trusty Grip Hoist. However in the ensuing years,so to speak, I'd learned a little of what I could do with this rock mover. S-o-=o-=o I decided that maybe I could do a little something and git a little gold if I did it just right.

After studying the situation a bit,I decided to follow a very small stream of water that seeped out from under the boulder pile. I figured this had to be the lowest point on the formation or the water would go elsewhere. On the assumption that gold would take that same route of travel, I started my operation. The first day was a day of rock movin but, I found a few nice finger pickers!! Hey this was pretty exciting and I kinda looked forward to the next day. Every day I picked up some pretty nice pieces, nothing great, but they started to add up. Our time was running out, so I had to be satisfied with what I had and pack it in. Added to the!!! major haul we'd made that season, I think we came away with ½ oz., most of it (the pretty pieces) coming from the White Marble Slab. Some day maybe, I'll continue where I left off.

One day before Thanksgiving, that year, we lost a dear friend, Delores, Elton's wife was killed in a car wreck. She had been a good friend and supporter of all we'd tried to do at Roaring Camp. We knew then that things could never be the same for us again. We pretty much gave notice the following season that we wished to terminate our employment.

We quit working the camp, the end of 98 and stayed on to entertain only. 99 was a pretty good year because high water ran out early, so in June, we took a look at things and decided to give that bedrock out crop across from Fountain of Youth creek a bloody go with the dredge. After some searchin, I found that out crop and we moved the dredge over there. The water was still almost two feet high in this area, but we managed to git her over there after a lot of pushin, pullin and swearin, etc. It was mid afternoon by the time we got under way, and Fran started cleaning the downstream side of that outcrop. There was a pocket like fault in the top and in that pocket, she found a nice little nugget! Well Jiminey!! That's pretty exciting, we hadn't been workin an hour yet and Bingo! We worked on some time longer, but the Bonanza was over,-No more gold. Well tomorrow's another day, we'll git right back to it.

WRONG!! The blame river dropped two feet over night and our dredge was sittin up on the rocks. Well that's a nice how do you do. We're all set for a good dig and the water run out on us. I spent a long two days pullin rocks trying to figure how to get water over there, no dice, there weren't no easy way to get there. It looked like we were gonna have to dredge from mid river over to open up that channel area.

Well, that's what we did, movin rocks and dredging way up till mid-August. We were two big boulders shy of makin it to our digging spot when we decided to quit. Early on we'd figured that there would be some gold we could pick up along the way, but no-o-o all we got was a little color and five weeks dredging with no gold show was a bit disheartening. I knew where we might could move to and do some good, so we moved up river to Slaughter House Gulch!!

Chapter 19

Slaughter House Gulch

Where do you reckon a name like that came from? Which liar am I supposed to believe? There's more than a few authentic?? Stories that brought about this infamous name. One that is fairly plausible, is; there was a mining claim registered in this canyon to an Alex Slaughterouse or the site of Alexander Slaughter's house. Another story is told, that the Miwock Indians resided in this canyon and were slaughtered by a band of Bronco Apaches. The story goes that the whole camp was killed, women, men and children. The canyon was thereafter called the Slaughter House of the Miwock's. This is a plausible story because originally this was the Calaveras River(River of Skulls). When Amador County was formed the river name became Molelumne, said to have been named after a Miwok Indian Chief. This last story though, is the one that I lean toward. It is said that in the early days, a slaughter house was located on the ridge above the canyon and that the owner's tossed bones, offal, hides and animal waste products down into the canyon, rather than waste time burying it. Knowing what human nature is, from being one of these creatures, I vote strongly for the latter story. Another reason,I make this choice is because even in this day and time,people still dump their trash there and have kinda made the upper end an unofficial dump site. It's also a good place to do a little shooting, target practice on the local rat population.

Moving on into Mining history, there was a forest fire (started by a kid smoking cigarettes) back around 1955. According to two resident eye witnesses, when the fire was out and all the brush was gone, the whole hillside was littered with holes. Some of these were twenty feet deep . All of those closely examined were dug down to bed rock and some even showed tunnels running off like spokes in a wagon wheel. This pretty well follows descriptions of other similar gold mining operations in Calif. The miner's dug straight down to bedrock and tunneled out in a wagon wheel fashion to clean bedrock for the highest productive values.

I was told that 6, 8 or even 10 miners would form a water co-op and buy water from the water company which ditched in water high up on the mountain. Now each of these miner's would dig all winter cleaning the rich values off bedrock and stockpiling this high grade near their Long Tom sluice boxes. Come Spring or Summer as required, they would all, upon time agreement, call for so much water to be released down hill in their prescribed and personally dug ditch. This water would provide the first miner enough to process all of his high grade and send it to a settling pond for release to the next miner down the hill. In this manner,each miner could process all of his high grade and they could all share the expense which,of course was a smart way to cut costs.

Well, now history is Ok, but we wanta git on with this here modern gold mining business. Fran and I moved into the river across from the Slaughter House Gulch Creek and started what was to become a major gold producer for a large number of recreational gold miner's.

This turned out to be a very good move for us, because, like I say, it started a Gold Rush that's still going on till this day. I had prospected this area some years earlier and it looked very good. I had sent people up there both to dredge and to Wiss Bang, but only one of the couples did any good. I was sure that with the right rock moving equipment,we could git some gold. Right—On!! We started picking up some nice ¼ and ½ dwt pieces right away. They were pretty well scattered,but made for a nice day every day. The accumulation after a few days of shallow dredging made up close to two ounces!

We had to pass by gold miner "Mockaloomie Jim's" cabin each day as we went to our home on the hill and he always came out to see how we'd done that day. Frequently we'd show him a few flakes and rattlers in the bottle and sometimes a clunker that we didn't dare shake for fear of breaking things. He had come up to visit on a few occasions and was always interested in what we were doing. His time and interest were always welcome and we shared lots of good times together. Especially, those Michael J's—That's Pepsi Cola and Black Velvet 50% each!!!! Yeah!! WOW!!

Fran and I were working one morning fairly close to shore and I had a 30 ft. suction hose on my super three inch dredge which helped us deliver the dirty water far out from where we were working. The only thing wrong with this was, it took a little extra effort and time to get back and forth . I was clearing a jam when I look over at Fran and there immediately behind her is the longest dang rattle snake I've ever seen. She is working in three foot water next to the river bank, and her back is toward the snake. Evidently,we communicate pretty good, because she looked my way, I pointed behind her and wiggled my finger pointing up and she immediately got the message. Rattle Snake!!! Now she don't see so good and to this day claims that the only reason she married me was because I looked twice as good (without her glasses). They were in a pouch on the dredge. This situation was cause for a little concern, so I took off immediately with my shovel and clobbered that big feller about six inches behind the head. there was four feet of snake coiled all over between these big boulders and about the time I'd half severed the body, this other head comes out from under the rocks.

My God!! I've got two of the beggers to deal with. That's where all of that snake body was coming from!

The one I had hit was trashing about trying to bite something, but he wasn't going anywhere. The second one tried to make a getaway, and I was able to pin his body against a rock. I held him there until Fran was able to shuck her dive mask, go to the dredge get her glasses and come back with our big pry bar and fully dispatch these two boogers. Now I'd hate to be accused of interfering with the love makin of any critter, but in this case I made an exception.

The excitement being somewhat over, I cut off the rattle's and put them in a large gold bottle. The head's I buried and the bodies I carried away for the local varmit's. Going home,that evening, I told Jim our usual, "Yea We got a couple of rattler's today, but I decided to give'em to you." He was somewhat taken aback at the prospect of some one giving away their gold, but recovered with a big smile when he saw what was in that gold bottle. We had to relate the whole story before gittin home and of course invite him over for a Michael "J" to kinda help us launch a couple of Rattler's to their Happy Hunting Grounds.

Well it looks like we're on to something, we keep dredging and the gold keeps a coming, no big nuggets,you understand,but real pretty pieces, plenty enough to keep us going. Then Fran goes down with a skin reaction to something that some how got dumped into the water upriver of us. We were working in a cove, not too much water flow, so the irritant kinda hung there on the water. We didn't notice it, so it got into her wet suit, causing a terrible rash and put us outta business for two weeks or better.

Aussie Sally and her family came to camp about then,so I put them to work dredging and having a good time. We moved a few big boulders, lots of little ones, dredged a bunch of river bottom and found very little gold. We really enjoyed the time with them, but our season was just about over. There weren't no more thrilling episodes to deal with and we packed her away for the season.

Chapter 20
The Common Operation

The boss man Elton, holds a Christmas Party every year for his company and all the folks that work for him. In 96, I had been looking at our Camp operation and recommended to him at that time, that we could put on at Roaring Camp a Common Operation that would pale all others. A common operation is where all the participants equally share in a mining operation and split the gold that's found. It was Christmas time, but, the idea took root and in the coming June of 97, he put together the first Roaring Camp Common Operation. It had all the bells and whistles that other's could not commit. We had High Banking, Dredging,Wiss Banging and Mill Trammel Operations. There were about 30 participants and all had a wonderful experience . We tried to cover all aspects of mining and provided a training and learning experience for everyone.

The proceeds were pretty good for all. Some of the operations were very productive and others more of a diversion from the intensity of feeding a sluice box. Our experience with the Common Operation led us to an enhanced operation that now produces the best results with the best processes and with the best benefit to our customers. Early on, we discovered that the best production of gold occurred when we followed the techniques used by the 49'ers. The Long Tom type of sluice box operation coupled with our Trammel output produced the largest amount of gold. The Wiss Bangr produces gold and is a wonderful recreational recovery unit for a small operation but, it cannot compete with the other units for cubic feet in processing. It's use is relegated to a diversionary gold mining instrument that provides a break from slugging buckets to feed a sluice box. Additionally, we try to perform the Wiss Bangr exercise in an area that is gold bearing with exploratory potential of a "Gold Strike" which could offer a big reward to all participants. The dredge requires experienced operators and productivity is very limited (this operation much too cumbersome and restrictive) for a common operation. Later years saw us abandoning the dredging as non productive and too restrictive for operation by first time gold miners.

The modifications that we made following the first few years, has finally been honed down to a common operation that everyone seems to enjoy. A large number of our first common customers are still coming back for more after 10 years of operation. We must be doing something right!

Many questions have come our way asking how we operate and how do we divide the proceeds? Our first concern is material, we must have rich and productive material that will produce gold. We have found two locations that we can count on for productivity and these two areas are where we get our material to feed the sluice box operation and the trammel exercise. Additionally, we prospect locations for setting up Wiss Bang operations during the common. We always hope for a Bonanza size nugget find or a large volume of gold for our operation. So far we have been more than satisfied overall. During one operation we accumulated over 7 ounces of mixed gold (based on averages reported) and during another we found 5 nuggets that topped 2 dwt. Additionally, we have seen multiple operations with ½ ounce plus nuggets. Our experiences have been unique and very memorable for most of our customers.

Well Tom! How do you divide 7 ounces or whatever of mixed gold between 30 people? This question always comes up and yes you could weigh it out in shares but, you have to understand; that weight alone is not a good indicator of what gold is worth. I'm always trying to illustrate to folks that gold is somewhat like horses. You see normal horses for the European market will bring 60 cents a pound, maybe a little more; among horse people, the winner of the Kentucky Derby, however will be priced a little higher. This is the same way with gold, nuggets have individual values that can go up to five times their market weight value. So, how do we manage to separate this gold and try to insure that every one goes home a winner.

Here is the process that we use. At the beginning of every common operation a Marshal and Deputy are selected from the participants. These folks are to bring to us, the administration; any issues that may be of concern during the week of exercise. This gives the members a simple avenue from within to voice any recommendation or otherwise that they may have. The Marshal or Deputy collect the gold from all operations and are welcome to observe the cleaning and handling throughout. All the gold from individual operations is put on display in a locked case accumulating daily for the weeks activity. On Saturday, once again in the presence of Marshal and Deputy, all concentrates are reduced and the fine golds added to the weeks gather.

On Saturday at noon all of the gold is placed in a gold pan for photos of the week's total gold accumulated during that particular common operation. Following the photo secession, I,an assistant, the Marshal, the Deputy and at least one other member of the common operation group take the gold to a separate room and dump the contents on to a large paper covered table. The paper has a pattern of small 2 inch circles that are numbered to match the number of participants in the group operation. I and my assistant separate the gold into 30 (or whatever number required) equal portions, in our judgment as near as possible to insure that each portion is of comparable value. Obviously, when a large nugget is involved, it is placed alone for the customer draw (which I'll explain shortly). Following my and assistant's separation of the gold I'll now ask for the Marshal and Deputy's recommendation for changing or reshaping the piles of gold to meet their satisfaction, I strongly remind them that they may draw the worst pile, so make darn sure there isn't one. When they are finished, I tell them once again that we, Roaring Camp; are now out of the picture and that they the common operation members are now in full control of the gold.

If they are satisfied with the divide, then we bottle the individual piles into 30 (or whatever) bottles. The bottles are placed in a row down the center of the table. If there are significant sized nuggets, these are placed in the center of the row.

Now comes the draw! Two decks of playing cards are selected with 30 matching values from each deck. One deck is given the Marshal and one deck is given the Deputy. All members of the common operation are summoned into the room . The marshal shuffles, asks for cuts and deals his deck face down, one card by each bottle. The Deputy shuffles, asks for cuts and deals all 30 face down on another table. All members draw one card from the table, the Marshal and Deputy get the last two remaining cards. The cards are then turned over starting with the end of row bottles. Matching cards get the matching bottle. This process continues until selection has reached the center bottles which hold the larger (if any) gold nuggets.

Commonly, some trading takes place between members when a special piece or nugget is or has a special value to an individual. And then the common is over, each member taking home a share of the gold—Next Question? When's the next Common and where do I sign up?

Chapter 21
"The Mama John"

Hey! you reckon that's an out house for the ol' lady—Well no-o-o, it's the name I gave to a special contraption that I put together so Fran didn't have to work so hard pullin the big boulders. You see in the past few years, since my shoulders gave me so much trouble, She's had to do most of that grip hoist operation and it takes a lot of energy. Since I ain't much of a hand at cookin, (with the exception of Micro waving), I felt it was kinda (if I wanted to eat) imperative that I take off the Heavy work load.

You have probably realized by now that I expend quite a bit of heavy thinking, trying to get outta work. This was no exception, I devised a plan to incorporate an air ram on a Grip Hoist and operate a control switch to work the arm back and forth. This was a beautiful winter project, because,at that time, I had to incorporate all of the dredge operation (water pump and breathing air) together with high pressure air all using one motor. I had a ten horse B&S motor at that time, so in order to keep my budget under control, that's what I used. I used the motor to drive a satellite set of pulley's which in turn drove a 250 gal per min. water pump, a twin cylinder oil bath compressor and a T80 breathing air pump. This unit was kinda bulky, but it sure did the job. In order to test my latest Rube Goldberg, I tied her off to a tree up at our motorhome and hooked my Jeep to the other end. That rascal pulled my Jeep up the road in gear and draggin all the way. All Right!!

Now I sat this contraption on the shore and ran a 2 in flex hose to the dredge, an air line to the Huka and an air line to the Mama John air ram switch.

It took some time to configure all this junk and get it operating properly, but I finally managed . The final outcome was really worth it,because it pulled rocks with no effort on our part and did it much,much faster. Fran was back to cookin again! We had started an operation late in the previous year at Slaughter House Gulch and had done very well. We wanted to go back and take up where we left off.

Well, Fran, sporting a brand new ball bearing BB weight belt went down where we thought was best and started dredging. Things were going great. We'd moved some big boy's using our new Mama John and dredging life looked good. Fran was down about nine feet when the wheels came off the wagon,so to speak. She Ran Out Of Air!! I'm sittin up on top,fat dumb and happy, everything looks good to me. What I didn't realize till later was, the pulley on the T-80 had come loose on the shaft. All things are spinning,but she ain't got no air!!

I may have said so earlier that Fran don't swim so good, about like a rock and her only way out is to walk. She was in bad trouble, but she didn't panic. She knew she had to shuck that 40 pound weight belt in order to get to the surface, so she popped the buckle, but didn't let it drop. She didn't want to leave her genuine brand new BB weight belt that cost $40.00 down nine feet cause she had no way to retrieve it. If she didn't git outta there pretty soon,she wasn't gonna need no weight belt!! This finally dawned on her and she quickly laid it up on a ledge and popped to the surface. I had no idea what was goin on till later. Her wet suit floated her,so she was no longer in any trouble .

From that point on,I made a few modifications to our modis oper-rendi and we got on with our dredging. We moved some big boulders with ol' Mama John and Fran enjoyed sittin in the shade operating that little air switch with every one . I'm not sure that she ever let me run the thing, mighta been afraid that I'd break it and put her back to work.

"Fran and the Mama John"

The following year,I devised a means of putting a Gast Twin cylinder Compressor unit on the dredge . This unit was a great improvement because I could use it to provide everything and it only required my 5 HP Honda motor. The operation did require pulling the suction hose out of the water to relieve water pump drag and provide max power for the Air Ram during rock moving. Additionally,it provided air for our under water air hammer operation. We had two of these air hammers that we used under water and they were the ticket for breaking up and cleaning that bottom of the river. Many times we looked back on previous operations where they would been mighty handy to have around.

Now, these hammers were the standard variety one can pick up at any hardware dealer but, the one thing important to their operation was to all ways run a little air tool oil through them at the end of each day.(if you wanted them to keep working!) One of our hammers was a miniature jack-hammer for breaking large bedrock formations and the other was much smaller for chipping out that small decomposing bedrock that always seemed to hide gold. Some folks feel that Fran and I go to some extreme measures in our gold mining efforts, but I strongly lean back on my Hillbilly heritage and that expression often used by my "Moonshiner" Uncle Robert, "You do whatever it takes to get the Coon!" On one occasion, Fran came upon a large 400 pounder she couldn't move and noticing that it had a small crack she used our trusty air hammer(which weighed 19 pounds by-the-way) to break it and easily rolled the two pieces outta the way.

Concerning the use of ol' Mama John, some of the biggest rocks we ever moved was with this unit and, (in memorium) we had it in use when we found our 2001, ¾ oz Pack Rat nugget. Roll on Big Mama!! Roll on!!!

About the time we had this Mama John working good. A three couple group came to camp from Victorville down near our winter hangout in Rosamond, California. They had a little dredge between them that they were gonna work and wanted to know where they should set up to find a little gold. We were prospectin up river a little ways at the time and told them where we had found some good stuff last year. Everyone and I mean everyone is skeptical when you tell them something like that, they immediately think why ain't you working there if it was that good. Well,Fran and I have always enjoyed the success of other folks, probably as much as our own. I've always felt that if you can't share what you have or enjoy with someone else, then it ain't worth havin.(Which,by the way is why this book was written!)

To make a week long vacation short, they enjoyed the best year ever finding gold. They were yelling and laughing the whole week,taking all kinda pictures underwater and just havin the time of their lives, They had never seen gold lying on the bottom before and were just thrilled to death. To say they were excited, it was the very best vacation they'd ever had and couldn't thank us enough. This kinda return for Fran and I lasts a lot longer than a few penny weight in a bottle, if you get what I mean.

Well, to quit reminiscing and git on with a gold story,it's necessary to say that we stayed in Slaughter House Gulch till 2001. Now the Fountain of Youth beckoned us again and as soon as high water went down, we started movin toward that little outcrop of bed rock . We had abandoned it two years earlier, but no one had seen the possibilities yet and we got to work. Bob Yawn and Gail Scott came into camp for a vacation and we had them join us for a few lessons in dredging and having fun. Gail became an expert at pullin straps off of big rocks laying on shore line, especially those that had a large iron shackle on the end that could hit her in the head. To her credit, she survived and will probably never make that mistake again. It really wasn't funny, but she's a good sport and laugh's about it to this day. We enjoyed a great week with them and although we didn't find a lotta gold, the strawberry margarita's were great along with that wonderful Blackberry Cobbler and ice cream that Fran made.

We had pretty well corralled that entry way by the time they left camp and only had a couple more boulders to pull before we could get into that outcrop we were headed for . Well, we shifted down stream and came in from the downstream side. We had lots of water flo now and we could dump our tailings behind us. This was a great move because Fran could see what she was doing. We moved in on that bedrock outcrop and two feet down we ran into deep disappointment. Nothin but decomposed granite, God! Had we ever had our time with this stuff. Every time we had come upon it there was nothing underneath, just bare bedrock all blasted clean.

Man what a disappointment, here we had such great expectations and we're met with our ol'nemesis Decomposed Granite .

We went ahead with our dredging, not really expecting a whole lot and we had run about one hour when we spotted a piece of gold down between the granite and the bedrock wall!! Hey!! That's a pretty nice piece!! Maybe ol'Mockaloomie Mama Nature, is gonna cough up a reward after all!! Now there was no stopping Fran,she started throwin out the debris and suckin up that decomposed granite like it was end of season!! She cleaned her down to about four feet and there uncovered a beautiful white seam of quartz covered with gravel and in the process of cleaning,almost threw away a yellow rock that weighed over ¾ oz GOLD!!! "The Pack Rat Nugget"!!

My God!! How could you do that?? Well, she didn't!! After recovering from the shock, she went on to get gold all through that quartz crevice, a final one day total of over 24 penny weight, This beating the 22 penny weight one day total, that I had pulled out so many many years ago on September Island. What a great end to a gold mining season, she thoroughly enjoyed and so did I. Her total for this operation topped three ounces.

One more season we worked ol'Mama John, and we started just below the Slaughter House Falls. We went down about ten feet in a pretty good size hole, moved some huge boulders to the middle of the river, one of which looked like a huge bear. We got one of Elton's old cowboy hats and glued it to the top of this rock, you could see it from camp . Got a lotta laughs but very little gold. Oh, Well, some things are more important anyway!!

About the time we decided to give up on this place, I remembered that the previous year, while finishing up that bedrock area where we got ol'Pack Rat, I had been standing out in the middle of the river to where I'd pulled the dredge . There at the end of the sluice box under water was a bedrock formation which dropped off into a hole where resided a fairly large boulder. By fairly large, I mean about four feet in diameter. Now this!! is my kinda country, where I'm pretty sure that no one before me has moved that rock or cleaned that hole. Compared to a lot of our efforts this was a piece of cake. It was just that, I pulled out the rocks and Fran cleaned out the hole. There laying on the bottom a small jewelry store collection spread out real pretty like for her. More book pieces for her and more great memories to look back on. Ol'Mama John comes thru one-more-time!!!

Later on,we purchased one of those back breaking 3HP driven 220lb. mechanical monsters, used it two years and spent about 8 weeks in shoulder therapy to heal strains caused from moving it. I've got to say that the ol'Mama John mighta been slow, but I sure came out better,physically, the years we used it . So here's to ol'Mama, may she rest in pieces. Some of which I'll no doubt use on my next, Let's Git Outta Work project.

Chapter 22
Bedrock Gold

Bedrock is described by many learned folks as the earth crust and as such it represents the bottom layer or lowest level of formation in any particular region of the world. In any mining area, this is commonly where the gold settles and remains waiting for some enterprising miner to get his hands on it. There are always exceptions to this common occurrence but I strongly lean in the bedrock direction. I am always looking for ways to get to bedrock whether it be dredging, sluicing, panning, or crevicing. This is normally the most probable location of gold and that's where I want to be. One of the most common problems I've run into is the fact that most exposed,easy to get to bedrock is on the cut side or outside bend of the river and unfortunately represents the wrong place to look because gold follows the inside or low pressure side. The outside bend bedrock has been uncovered and washed hard by river action cutting away and polishing the surface thereby preventing the formation of gold deposit areas. Having said all this, there's that once in a decade goof-ups in gold mining when you can break all the rules and still come out a winner.

I was High Banker Wiss Banging just below the China Camp Falls on the South Fork of the Mokelumne River. The water was too high and much too cold for an old sunshine dredger like myself, so I had deferred my water activities to later on in the summer. The formation of bedrock in this area looked kinda interesting,but it was on the wrong side of the river for good gold deposit. Every time I looked up river, it still looked bad,but there was no way to cross over to what I considered to be the more promising area. I had to set up on this near outside bend of the river side and hope for Mama Nature's Compassion. I had explored other more promising areas on my side of the river,but the work involved in getting my equipment to these places represented an effort that I just kind of shied away from. I was kinda looking for a place to work for a couple of weeks until high water receded and things warmed up.

The interesting formation of bedrock that got my attention was a huge basin like depression in this riverside that looked like a natural trap if you could imagine it without all the boulders and deposit material in it. It's true that it was on the wrong side of the river and looked like billions of tons of water had worked it cutting out it's center and making it a virtual bowl. The deposit material was an accumulation built up over years from slow moving receding waters. Basin and all,it still didn't look too promising. One major thing captured my interest,and that was a crevice on the upper sidewall of bedrock that appeared to go down toward the bottom of the basin. Right in the middle of this hopeful glory hole was a big old boulder. My shoulders started giving me trouble just looking at it.

Nothing ever comes easy,so we better get moving,literally; if we're going to clean this place out. I've progressed to a small hydraulic winch and motor to pull my boulders as certain physical inabilities have caught up with me. This unit enables me to rig up and move almost anything we come upon.(Of course,it helps to have a wife with a strong back!) The winch must always be situated to protect the operator from the whipping of a possible broken cable. When you set your rigging, always insure that the hardest pull is always on the dead man and the load, and never place yourself or the winch which you have to operate, in line with these heavy pulls.

Keep yourself safe at any cost and be in a position to be a spectator in the event of a break; not a part of the action. I always pull from an angle outside the line of fire. You know, if you find a little gold, you might want to be around to enjoy it.

Following a week of Wiss Banging, we've cleaned out the basin and found enough gold to convince ourselves that some came down this side of the river. Mama Nature sometimes takes pity on dumb ol' gold miners when they can't cross the river. (Not often enough!) Our bottle is really looking surprisingly good (about 1 dwt.) and Fran and I are neither too awfully disappointed that we can't get in that c-o-o-old water to do our dredging. We'll get there later on but, right now it's time to clean out that side wall crevice. I've scraped, scratched and washed it out as best I could,but it's now time to make things even easier for this ol' gold miner. A feller just has to use every little labor saving device that's handy.

I brought in my trusty air hammers to lend their valuable assistance in cleaning out this potential Bonanza. These air hammers are,in this instance powered by my portable air compressor unit which I'll get into a little later on. Anyone having seen their commercial use in everyday tasks will recognize their potential effectiveness in opening up a rock sidewall. At this point,I have to mention(once again) that air hammers will do a great job for you if you remember to run a little oil thru them after each day's use. I've lost a few when I overlooked this important little requirement after using them under water.

The large boulder is outta there and the bottom of the crevice is finally open, my poor ol' back is hurting, we're about outta gas, Fran is tuckered out, and we're getting ready to call it a day, When "THERE IT IS!" I don't believe it either! A pocket of gold as yellow as it ever gets! My God! What a payday, What a Bonanza! What a great ending! It's been laying here for thousands of years just waiting for us! But Lord! Would you just look at That!

There were no big nuggets, but days later when we had recovered and cleaned it all; It made about the best 6 ounce Happy Face I ever saw!

"How's that grab you for over six ounces?"

Chapter 23

Horsetail Gold

The Horsetail is a prehistoric type of plant that grows in sandy soil usually very close to a water supply. It's uses are many as it has a high silica content, that allegedly can be of some medicinal value,but it's primary function during the gold rush days was for washing pots and pans. It's silica sand abrasive characteristics made it a natural scrubbing and polishing device, as the long stemmed plant, could be folded up and bundled for use as a brush. During the 1849 era, a majority of cooking was done over an open fire pit and the utensils were always in need of a little Horsetail attention. Fortunately, my association with this plant had nothing to do with either of these uses.

My introduction came during my younger years on the farm; when I was assigned to eradicate this Horsetail weed because it was a slow acting poison for dairy cows . It was very difficult, almost impossible to get rid of and I learned a good deal about where and why it grew in our cow pasture. The roots would always come back the following year and I had an unwelcome familiarity with the nature of this beastly weed. I never looked forward to spring because I knew ol' Horsetail and that 5 pound Grubbing Hoe was waiting for me. As a ten year old, I figured that any cow dumb enough to eat this stuff wasn't worth having anyhow.

So where does the Horsetail fit in gold mining? Hang on a little bit longer, because this is that story.

The location is on the north fork of the Mokeloumne River a little ways upriver of Camp.

I'm planning on mining this large dry bar (a long deposit of combined river material that serves as a barrier to water flow) down stream of a bend in the river. This bar is on the inside bend and the perfect drop zone for gold. My problem is, this is a huge bar approximately 250 to 300 yards long by up to 100 yards wide and I'm looking to High Bank a small portion of this megatons of material. Naturally I expect to dig in an area that will produce the most gold . Yeah, right on! So where do I start?

I probably spent the better part of three days prospecting that bar to try and find some indication of where a possible high banking site might produce some gold. I found some surface hardware and panned out a few prospective areas that produced a good quantity of black sand, small garnet stones and some fine gold. This, however was all surface deposits having been dropped during recent floods and I knew from previous experience dredging the adjoining river, that bed rock could be 15 feet down. I also knew that, normally; I had to get to bedrock before any larger gold showed and of course, I had to be in the right place. This bar only flooded over with hard moving water every 25 years and then only for short peroids, so any deposits were going to stay close to the surface and any really big gold was going to be that which was deposited during the thousands of years past. I kept looking for that something that would possibly put me in a good hole. In an expanse of this size, I'm looking for a needle in a haystack.

Well sometimes blind luck beats all the theorizing and gold mining skills available. I have no idea what that bar looks like under this expanse of rocks and rubble and was thinking what a quantum size problem I've cut out for myself. This bar is a no brainer, go somewhere else! Surely this river has thousands of places where a high banker can get a little gold without moving half of Texas. Well, optimism is a major affliction of all gold miners, so contrary to any logical reasoning, I'm gonna stay right there until I have my shot at mining this bar.

For whatever reason, I was attracted to scattered patches of foliage growth, maybe because they were the only things standing out on this bar. There were a number of areas out there where dry grasses and berry vines and some weeds were struggling along; then I saw the Horsetail! It's growing right out in the middle of this dry expanse where most everything else has dried up and I knew from my youthful experience; it has to be practically sitting in water to be growing green and healthy this time of summer. How can this be? There has to be a good reason and I figured bed rock has to surface shallow underneath, or a channel comes in from the river; to provide the water necessary for Mr. Horsetail to survive. This also means that this shelf or channel possibly has the characteristics necessary to provide a perfect drop off and hiding place for gold when the river flowed thousands of years ago. It is now High Banking Time! This is a spot to check out!

I'm going to be working in the open sunlight and in the normal (now dry) river channel, so I assemble my high banker Wiss Bangr and proceed to dig a hole. The flusher fogger spray nozzle allows me to wet down the whole area including me. This cooling effect makes working out there a whole lot more enjoyable. The usual number of boulders are blocking my progress, so they had to be dealt with first and then I was able to begin my exploration.

I couldn't get much luckier! Four and a half feet down, I hit the first sign of bedrock! Boy, you talk about happy,this is really something! I'm really getting excited now because the bedrock that I see is ancient and decaying and on top of all that here in my sluice box is a little bird shot size nugget! Man it just don't get much better than this; what with the little work I've done and the rewards already starting to come through; I just can't hardly believe it! I couldn't wait,

I just had to pan out that box and sure enough, there were two more smaller pieces with a few flakes and fine gold. This could be "The Glory Hole" I've been searching for! I had to get back to digging! Big stuff is laying there, just waiting for me! Hold Your Breath! Black decaying bedrock is showing on the left side of my glory hole! Boy's, I need to get to digging!!

But Whoa Up Tom, ol'buddy!! Sometimes Mama Nature is an unforgiving lady that doesn't tolerate oversight on the part of ol'gold miners. I had over looked the presence of red dry leaves in the area plainly spelling out "Poison Oak." I had dug right through roots of a bunch of this stuff and was gonna pay the price. My arms were in trouble and my enthusiasm for continuing to dig was severely dampened. Fortunately, I'm not allergic to Poison Oak, but I sure don't like the discomfort, even if it's only for a couple of weeks. My mining was put on hold for a bit.

I'm still highly excited about the prospects showing in my diggings, but I had to get somewhat well before continuing. Perspiration and Poison Oak just don't mix. It's probably a bad choice of words but, we're going to scratch out my time off and get back to gold mining. I waited as long as misery forced me to but,the lure of that little piece of bedrock had me back out there way before old poison oak was gone.

That hole grew fast. It's amazing how much work you can get out of a Gold Fevered old gold miner. The boulders just seemed to help me get them out of the way and a little gold kept showing in my sluice box. If I'd slowed down enough, I would have realized that my sore back and aching muscles didn't think them rocks moved themselves at all. My bottle was really looking good and the bedrock was opening up into one of the prettiest text book crevices you ever looked upon. I and Fran have opened up some great ones underwater, but nothing better than what we were looking at here. This one is running at right angles to the water flow, jammed full of small gravel and sitting slap dab in decomposing bedrock. The prospects for finding some nice pieces of gold have never been better. Boy,you can almost smell the gold!!

I had to slow down, as there was a huge rock teetering on the edge of my Glory Hole and gold fever or not, this big bad boy had to go before I spent more time in this hole. I had been advised by good friends, to write my name on it just so the spelling would be correct on my headstone, (If it fell.) Nice Advice! Huh?? So, out comes the Grip Hoist rock puller with all the cables, blocks and shackles and I start laying out an extended dead man ½ inch cable to get out to my working area. I rigged it up for a downhill pull over and past the crevice. It was about five feet in size and would move out with maybe three blocks. I figured to be back breaking open that crevice by the end of the next day. Wrong!

Disaster strikes! Not an accident or injury but for a gold miner just as bad. Up river at the big Salt Springs holding reservoir, a crack had appeared in the side wall. The reservoir had to be drained for repair and all that water was coming downriver, onto the bar and flooding out my glory hole. There was no amount of studying or planning to do .Get your equipment out and get it out fast! Two days later here she comes, two feet over the bar and rolling hard downstream. There goes my glory hole and there goes ol'Horsetail. A disappointing end to great expectations. A Flood! It covered up everything! Good bye Glory Hole! Good bye Gold and Good bye Vacation!!

Well anyone that's been gold mining for any time, knows the meaning of adversity, bad luck and the fickle personality of Mother Nature.So count your loses, bail out, check your gold bottle (Not too Bad ½ oz!)

and by the way; before we get too down hearted, remember and be thankful for the nasty persistence of ol'Horsetail. She'll be there next year growing green, looking mean, and pointing the way back to that "Glory Hole"!

She was and I was(there thc ncxt year) but other circumstances pulled me to other new adventures and besides, I knew that I could always come back if things didn't "Pan Out" elsewhere.

Chapter 24
Bowl of Gold

How do you separate black sand from gold? Well this is one of the toughest tasks facing a gold miner. There are many devices and techniques out there that do the job. Some folks use mercury amalgamation and do a great job; it's kind of risky for those that don't know how to properly handle mercury and it can be very dangerous for the novice user. There's panning,of course; and I'll admit I've seen some real experts like ol'Jackson Neal of Pine Grove and Jim Lewis of Oregon that really separate down to the fines in just two minutes giving you the impression that it's easy. Wrong!! These guys have thirty years of experience and it's definitely not an easy task. Try it and you'll probably lose half your fines plus a lot of your valuable mining time.

Then of course there's the gold wheel, a good dozen varieties and the spiraling gravitational action will separate the gold from the black sands, if you've got the time and patience to monitor it and the water supply to get an optimal separation. It does a commendable job for some folks . Other's feel it's too slow and opt for something faster. Well, there's the micro sluice which does a great job and commutes a larger volume of material in the same time however, there's still the separation from some black sand necessary. Next we deal with the shaker table which does do a pretty wonderful job and it's side kick the slate table with squeegees and a definite ability to capture that small gold. The biggest problem here is size, manpower time and cost; they're definitely something for the commercial miner as are flotation tanks, electrolysis, leaching chemical recovery and furnace processing.

If we drop back to the very early practices of black sand /gold separation, we find that it became a woman's job(if there was a woman around). When panning techniques had reduced the concentrates to about a 50/50 mixture of black sand and gold; the next procedure was to dry the material and take a lock of hair and make a small brush which was used to spread the material on a flat surface and gently blow the black sand and small garnet stone free of the gold. The fine gold could then be brushed onto a corn husk or piece of paper for emptying into a leather bag.

Now these are all tried and true means of separating black sand from gold, but for the recreational gold miner, there is a definite problem that I have encountered. You see, I have a wife(with beautiful hair), a gold wheel, a micro sluice, a mercury retort unit, twenty ounces of mercury, a bottle of nitric and another of sulfuric acid and numerous types of gold pans. I also have occasion to use the shaker, slate tables and smelting reduction to get my gold. Some of these required use of friends equipment and as I am hesitant to impose, the problem had become thirty five gallons of rich black sand in my garden shed that was steadily accumulating and I had not found a satisfactory lazy man's, time saving, means of processing it. That was when I came in possession of the famous D.A.M. Gravity Concentrator Bowl.

Well, where did this thing come from? I don't know that I or anyone else has the true complete story, but I did run into and now own an old dairy cream separator which is suspiciously like the concentrator bowl. The water/gravity method of gold separation interested me,because: some years ago, I was shown a water concentrating device. It was a five foot tube of glass with a funnel in the top. Water was injected into the bottom. Material (black sand and gold) was introduced in the top (probably a cup full). The water was turned on and adjusted to per bate the material such that the black sands were driven to the top and over the edge of the funnel. The gold settled to the bottom and could be recovered in this manner. Major heavies remained with the gold but this method did get the job done. How efficient was it? I have no idea, but the principal seemed OK. The rest speaks for itself as transporting a five foot glass tube would get to be a problem. Following this introduction to gravity separation, I heard there was a commercial mine working in Canada that used an electric centrifugal cream separator for doing the job. I would suggest that this could be possible with some modifications to the unit, but; once again I'd question the efficiency, the size of the operation and the cost involved.

Getting back to this Gravity Bowl, my wife Fran and I were hitting garage sales on Saturdays a number of years back when I came upon this funny looking pot for sale. It had an angel food cake pan look with a hose coming in from the side. Upon closer inspection, I knew that this was a gold separating contraption that I'd heard of somewhere. I really had no idea how it worked and the folks selling it didn't even know what it was. I thought a feller could risk a dollar on this thing so I offered them fifty cents. They took my money,glad to get rid of this angel food pre-boiler thing. Little did I know just what I really had come into. This was a genuine gold gravity concentrating bowl, a very early model of the easiest to use, most efficient and cheapest black sand/gold separator on the market.

I took one close look at the way it was built and said aloud, "I bet some wife was really upset when she saw what her gold mining husband had done to her beautiful angel food pan!" Next question, Does it work? Well first,I gotta find someone to show me how to operate this thing . I did and believe me the bowl did also! Was this really the ticket for running my black sands and separating the gold? I was told that among old prospectors, this bowl was well known and most of them had one or knew someone that did.

My first trial involved using the readily available garden hose. Bad Move! I had the whole area flooded before I even got started. This immediately called for a small bilge pump and a means of re-circulating the water for continued uninterrupted use. Now we really had something going and if this thing worked, we could get serious about processing my mega-buckets of sand.

My initial efforts and results could have been a lot better, if I had followed practices that I heard of later or learned by trial and error on my own. This little gadget worked to develop a water vortex that lifted the black sand up the sides of the funnel like cone in the middle of the pan, over the top and exit into a bucket below. Yes! It worked great as long as the particles in the bowl were not too big. A size 20 or smaller screened material worked great, but I found that some very fines would be washed out with heavy water action. Like all equipments, some degree of proper use is necessary to get good results.

Did I get good results? You better believe I did! This thing worked great; a tremendous improvement over anything I'd tried before. I started screening my material down to 20 size even before I brought it home. I'll admit my early efforts were fairly slow and cumbersome putting together a pump and catch basin arrangement but, as I learned a few short cuts; things really smoothed out for me. Examination of my overflow of black sands indicated a very low loss of flour gold. If I didn't "horse" the water supply, this little bowl worked great. So what happened to this Gravity Bowl? Well technology and time and the rising price of gold caused some major changes over the past 15 years and now we have "The Blue Bowl," a much improved version made out of plastic. Today's model is manufactured in large quantity and readily available through most

Gold equipment suppliers. I can sure vouch for it's benefit to black sand/gold separation and your rose bushes will thank you for that throw away black sand. The wife will no doubt appreciate getting rid of all those rusty coffee cans and buckets cluttering up the garage. In my case about 1 ½ oz of fine gold is a whole lot easier to handle than all that black sand.

The following set of operating instructions represent the tried and true methods that have worked for me. First, don't even bother to pan out that sluice box be it from high banking, straight sluicing, dry washing or dredging. Now right away you're gonna say why Tom, that's crazy, but I'm here to tell you that during any panning operation you're gonna lose some gold. Those microscopic pieces will leave your pan like the gates are open, you'll never know they were in there. The surface tension of water will pick up that fine gold and good bye. What I'm recommending is a process that worked best for me. I water screen all of my sluice box material through a ¼ in. screen and follow with a 20 mesh. The contents of the ¼ in and the 20 mesh I pan out and pick out those nice bigger than 20 mesh pieces. They look wonderful in a bottle. It is now time to run the sub 20 mesh in the Blue Bowl. The manufacturer recommends screening to a 30 mesh and he may be more correct than I am but, 20 mesh works for me . I do a water feed rate at 25 percent over flow on what I call about 300 degrees from the jet inlet. I try to get the bowl perfectly level as any miss-level really reduces the effectiveness of separation. This permits the fastest quantifying of the sub 20 mesh material. All of the micro dirt and white sand will be overflowed into the funnel along with a large portion of the micro black sand. My operations usually involve 1 gallon of 20 mesh screened material . This usually requires about one hour of running time to get optimal results from straight material.

At this point,my water is so dirty I can't see the bottom of the bowl, so I pump off all of my dirty water and get a fresh clean supply.

The material in my bowl at this point appears to be mostly black sand with some gold showing around the edges on the bottom. I can take the large water squirter (turkey baser) and per-ta-bate (it's a new word which means to mess up the mind) the material piled up around the cone of the bowl. This permits a dynamic water classification action which gets rid of more small particle black sand. This activity is kinda fun but very slow to clean out all that heavy black and finally get to the gold. Most of the content in the bowl at this time is magnetite (heavy black sand), small river garnet stones(they have a higher specific gravity than other river rock material) and hopefully a good amount of gold.

Where we go from here is a process that, once again; works for me . Your methods or the manufacturer's may be better, but so far I haven't found any. I take a fairly strong magnet, place a plastic baggie over it and rest it on the top edge of the funnel in the Blue Bowl on the edge opposite the water feed. A gentle per bate of the black sands will cause them to elevate in the water and adhere to the magnet. The word here is gentle in order to prevent the pinching of gold between the magnetite. Should there be any question, simply take the bag from the magnet reversing it inside out and now you have your black sand in a baggie ready for inspection in a gold pan.

Now,take a look at the contents of your Blue Bowl. Notice the red garnet (river rubies)color of those little mostly round stones in the bottom with your gold. At this point,most folks will slow down the water feed, per-ta-bate the material to drive the gold to the outer bowl perimeter and take a snuffer bottle and pick up the gold. This process will work fine,but I prefer to snuffer up the garnets and other material that migrates back to the cone first. When I have all of the unwanted material in my snuffer bottle, I run this thru a 50 to 100 mesh screen, in the event that I may have snuffed up some micro fines in the process that may have run in with the garnet. The final process involves getting all of your gold out of the Blue Bowl without dumping the water and a snuffer bottle will get it all at this point. I have noted in the past that some micro fines will ride the surface even with soap in the water when the vessel is tipped over. Those micro fines are like talcum powder and you can lose them in a heart beat if you're not careful, I have found them sticking to my fingers and retained in the snuffer bottle tube. They can be very nasty to deal with, so I always acquire and move them underwater and finally permit evaporation to settle and dry them.

All of the sub 20 material has been retained in your catch bucket under the Blue Bowl and now it can be screened down to 100 mesh and rerun with a slow water feed to insure capture of all those micro fines that maybe got lost during the 20 mesh run. Using this process is probably the fastest and most economical,for the beginning gold miner; without any risk involved with mercury or the expense of other more sophisticated operations. It's a great winter project and it works for me!

My last run of the previous year dredging produced three and a half gallons of black sand. The Blue Bowl turned out one quarter ounce in fines from that batch.

Try it, you'll not go back to anything else!

Chapter 25
Falconing for Gold

Does this title get your attention? You think maybe it's got something to do with a bird or fowl picking up gold nuggets. Well, the only time I ever heard of such a thing was in the late 50's. I was spending my week ends working for ol' friend John Lee Rymer, in Frazier Park Ca. and he was well acquainted with the Goldstine family and their turkey ranch up the road towards Mt. Pinos. John was telling me that every turkey crop they raised and sent to the meat packers in Newhall, was sold with a reservation; all the bird gizzards were to be packaged and returned to them uncleaned. Why? Early in their attempt to start a turkey ranch, they slaughtered and cleaned a number of the birds themselves, to give to their neighbors. They discovered that some of the gizzards contained small gold nuggets.

Well, I was not much into gold at that time, but to me it was kind of an amazing and wonderful thing happening to friends of ours. John Lee was from Monroe County in Tennessee and said that he had heard of folks finding gold in chicken gizzards up around the little town of Vonore and even to this day, every chicken in town has to beware of hatchet carrying gold prospectors. Turkeys and chickens consume small pebbles and retain them in their craw or gizzard to grind up their food and I reckon gold fits that description.

The Goldstein ranch turkey grounds was rotated in two acre plots and somewhere in one of these, there had to be an outcrop of some kind that produced gold. John and I went and had a look at this place even though we knew nothing of gold mining and of course, we found nothing. To date,I don't know that anyone has been able to locate the source of that gold. Oh, and by the way, this turkey/chicken gizzard bit don't have anything to do with this story.

I was first introduced to a little metal detector "The Falcon" at a Gold Show sometime around the early 90's . Now, I'm not sponsoring or selling this instrument, but I am saying that for a lazy ol' gold miner,like myself, it sure makes the job of crevicing a whole lot more productive. The technique is really very simple. The head of the Falcon detector is close to one inch in diameter and designed for small area scanning. You scan small shallow crevices on large margins of exposed bedrock trying to detect a signal. The Falcon requires you to be very selective in the area's you work and permits scanning of those very difficult area's not open for normal size detector heads. Frequently the sidewalls of depressions and crevices are very narrow; great for catching gold but difficult to impossible for most normal metal detector scanning.(Believe me, I've tried it with a number of makes and models.

The latest Model "Falcon Metal Detector" is great for crevicing wet or dry!

The Falcon comes with a test card that has a flake of gold attached to it for test purposes. Upon first seeing this test nugget? I considered it a disgrace to call this miniature flake a piece of gold. It very closely falls in the category of a piece of fine gold. At the time,I thought the manufacturer could have given me a respectable piece to test the instrument with. Quickly I found out, that this little mite of gold, was plenty big enough for calibrating the instrument. I even turned the card over and added material separation but,this little guy performed admirably. I received a loud beep for every pass that I made.

Now, after being duely impressed, I had visions of hundreds of square feet of bed rock, that I knew of, in some terribly rough areas that I could scan with this baby. Naturally I had to have one so I purchased it for my wife! She's not much into crevicing, but since I'm such a nice guy; I'll only use it when she's otherwise occupied. This little mis-adventure turned out to be a real blessing in disguise, the detector was a real god-send for finding near surface gold in those little hard to get to crevices. I almost gave up speculating on some when I had no reading at all. I did pass up a number of good looking ones when my mining time was short. The real answer came though, when I got a signal, there was always something there. It maybe was a black sand pocket, a buck shot, a fishing weight, a piece of metal, part of a beer can or a piece of gold! Exciting? You bet and very rewarding.

On one occasion, I was headed up river, when I come upon a young couple (Kim and Eric of San Diego) working a bed rock sidewall in an old mining excavation known as John's Hole. It was early in the year, mid June and there was considerable water to deal with. They asked my opinion on their operation and I recommended a dry area further up stream as the hole they had in bed rock, was full of water. Well, the young lady implored her husband to stay there as she felt there was gold in that area and she asked if there was any way to check the hole out before they gave it up. To appease her and get her off her husband's back because he wanted to move, I suggested checking it out with my little detector because it was waterproof and if there was no signal, then a move was obvious. Guess what? That little rascal went off like a three alarm fire; it sounded like a ¼ inch steel washer in that hole. That or something similar,is what I thought it was.

That gal used a brand new beach towel to soak up and remove all the water from that hole. she dug out the bottom and panned it beach towel material and all. When Kim yelled out, "What color is gold!!??" Eric yelled back, "Yellow, like butter!" There it was!! A nugget!! Lordy! that lady was speechless. Her husband couldn't believe "that" either. Right On! It Looked Just Like Butter—3dwt+

Now, Eric was a confirmed die hard dredger, who didn't wish to waste his time digging shoreline bedrock, but to this day their biggest find is that nugget that came from John's Hole weighing over three penny weight. I reckon we'll have to put another star up on the wall for women's intuition!

Now, I'm a gold miner that wants all the bells and whistles and every piece of equipment available that will "get the gold." Assuming,of course that I have the time, the money, and the opportunity for using it. This does get to be a problem, but I managed to get the time to use this detector and get it into some very productive area's that really produced the gold . Yes, it's like the man said; "Butter is the Color of Gold." So-o-o-o, "Let's hope that all of your gold mining is loaded with the color, butter!" Mine was! It registered over two penny weight found with the Falcon!

OH! Yes! One last thing, keep on the look out for "Homeless" chickens in gold country, they make excellent dumplings. I had my an eye on a vagrant young rooster called "Barred Rock Bonanza," seen pecking around on Little Rich Bar down river. Naturally, my only interest was to save him from some Bobcat! Unfortunately, someone or something got to him before I did! I do kinda wonder what his dwt. was!

Chapter 26
Let's Go Crevicing for Gold

At the risk of turning this story into a training secession, I have to hold up on these thrilling episodes to lay a little education on you.

The use of the Falcon metal detector reminds me that a lot of folk's don't have access to one and depend on older methods for gold seeking. Crevicing is one of the methods I personally enjoy as it doesn't require a lot of investment or labor intensive hard work. The following is a description of the techniques and methods that have been successful for me.

Folks have been looking for gold in every manner possible since the big gold boom in 1849, so what chance do we have of getting anything in todays world where literally thousands of prospectors and fortune seekers have been before us. To answer that, almost every river in California is gold bearing and has been thoroughly explored, but as I myself can attest to, nobody ever gets all the gold. There are virtually millions of gold deposits still undiscovered along the banks of these rivers. I have been prospecting these possibilities for the past 26 years and have been more than moderately successful High banking, Dredging and last but not least Crevicing. This is a technique of gold hunting that involves cleaning out crevices and malformations in the bedrock and large boulders that serve as the stream banks that channel our water ways through out the southwest.

Where do we find these gold bearing crevices? Well like the man said, if I knew which horse was gonna win the race, I'd bet the farm and if I knew which crevice had the gold, that's where I'd be diggin. So instead of having all the answers, we'll try to give ourselves the best opportunity for success, by following the best practices and principals available for any placer or free gold mining operation. Every book written on Placer Gold Mining, tells you that the inside bends of rivers and streams offer the best possibilities for gold deposits and looking for rich crevices is no different. If we take a look down river in this particular location shown in the following photo, we see the river channel is lined with possibilities. Every rock and outcrop is a possible location for gold deposits . The river bends to the left, so the most likely location would be across and downstream of where I'm standing. Well, my usual efforts involve trying to decide where to start this crevice cleaning, or maybe, instead just sit on a rock and watch that ol'river run. You gotta remember that gold mining always involves work, so let's not get in a rush; that gold ain't going anywhere!

Like all gold seekers, I have always had the feeling that unless I'm moving cubic yards of material, I'm not gonna get much gold. Sometimes old age and infirmaries force a lazy man to seek easier avenues and start using other methods for gaining rewards. It really doesn't take a lotta gold to make a feller plumb happy. For instance, take a look at this picture of a tablespoon. This is my genuine silver plated brass spoon that I have hammered into many shapes to gain entry into a crevice or small hole in bedrock to clean out the contents. Fill it to the brim with gold and Man What a Payday!!

This particular spoon, by the way, owns the unique distinction of having extracted three times as much gold from crevices and potholes than it's own weight. This statement can be confirmed by my daughter, Loreea, Butch and Nancy Kirkwood of Kansas and "Lost Fred" of the Lost Fred Mining Company of Ohio (whom wanted to buy or steal it for a trophy). Their individual successes, they still attribute to this Silver Spoon. When I loaned it out, my only advice to all of them was, find crevices and pot holes in which it will barely fit and clean those out. They did just that and behold, they each came home a winner.

First let me say that what we're looking for is not those crevices, cracks and declivities that would permit a shovel, scoop or crevice tool entry. Anything that big has been hit by at least 200 gold miners before you. Let's set our sights on those little three or four inch crevices and cracks that are only a quarter inch wide. These are too small for any self respecting gold miner to fool with; therefore they're perfect for our technique. Remember, we're looking on the down stream side of these formations and large boulders that have been sitting there for thousands of years with small openings that permit the capture of gold.

These little crevices, on one of my morning excursions produced over a quarter ounce of nice pieces of gold. The whole story behind this excursion was not a ringing success, because my wife, Fran ran a long way down river to find out what happened to her lost (possibly injured) husband that had been sorta delayed from an important appointment by "Gold Fever!!!" She was not a happy camper for no amount of gold, I would have been better off maybe if I'd had a broken leg .

OK so now we take a close look at just what I'm talking about. These small cavities are formed where differences in rock formation occur and soft textured rock formation wears or decays away at a rapid rate. They form perfect pockets for gold deposits and offer a treasure trove for the recreational gold prospector that wants to go low budget and light weight while still enjoying that quest for gold.

Small crevices located on the down stream side of bedrock formations frequently only produce a quantity of fine gold, however; there are literally thousands of these that have never been touched and one may be waiting for you that hides a very nice nugget. My experience has been "You find a hot spot and almost every crevice holds flakes and pieces of gold."

So besides lots of water and a lunch, what do we need to go on this Glory Hole Expedition? First of all let's get us a water purifier to carry, water weighs about a pound to the pint and a gallon for the day might be necessary. Believe me, the water purifier is much lighter and perfectly safe. Next, let's take a look at my bag of tricks.

"My Bag of Tricks for "Gittin the Gold!"

The necessary tools are as follows, a large handkerchief, a short handled 1 lb mall and a couple of chisels for breaking open and breaking up the contents of a likely crevice, a long thin flat blade screwdriver, tweezers large 10 inch and small 4 to 6 inch, two halves of the plastic lid of a three pound coffee can and a 1 inch & 3/8 inch paint brush, a silver plated brass table spoon (optional for good luck), a two foot length of plastic tubing with a 1/8 in. inside opening, a ladies compact mirror (2 inch diameter is very good and housed in a plastic case for protection), a 6 inch dental cleaning tool with sharp "L" shaped points. Finally,of course you need a gold pan (plastic preferred Roaring Camp—light weight, for best performance and heat reduction) and a gold vial with medicine bottle carrier for break protection to hold your findings.

After reading this, you begin to wonder if this guy is looney or something; whoever heard of gold mining with handkerchiefs, tubes, mirrors, compacts and plastic lids. All I can say is hang on a little longer, I really haven't been out in the sun too long.

We finally have all this good stuff and so we set off down river to try our luck. I've traveled this river many times but, I still find hundreds of places that I missed. There's one thing about rivers and rich deposit areas, the gold continues to be deposited in the same areas until time and physics change the way the river flows. It's true that sometimes this takes thousands of years; however, there's also those areas where some replenishment occurs almost every year during high water periods. The photo's shown are true crevices found in the earlier river picture. I'll admit that on this particular day my luck ran pretty thin and I did discard a numberof non producing crevices before exploring one that was a good producer.

Notice! It's location was on the down stream back side of a bed rock formation. Using the mall and chisel, we open up the crevice and attempt to break up the content to it's bottom. Pick out the larger pieces with tweezers, then folding the ½ coffee can lid and inserting it into the crevice, sweep the contents into it with the paint brush and dump it into the gold pan.(Note: The plastic ½ lid will take any form and offers a good pick up scoop for any small crevice)

Continue cleaning in this manner until it appears that the crevice is empty. Take the compact mirror and use it to reflect the sunlight down into the bottom of the crevice. This bright spot will usually plainly show that more work is needed. Frequently compacted gravel and dirt still remain in the crevice. It's now time for finer detail cleaning with the small chisel, the screwdriver, and dental tool (remember the best gold is usually at the bottom of the crevice on bed rock). It's always a problem attempting to get all of the material out of the crevice so here is where that plastic tube comes into play.

Take the tube and use it to gently blow the material from one side of the crevice to the other (gold pieces, even though small, will remain in place), and once again using the compact mirror, illuminate the dark bottom to insure that you have thoroughly cleaned the crevice. Many times this last examination has shown me the gold just waiting for me to pick it up. It has been my experience that thin flat pieces of gold are usually found on the crevice sidewalls standing vertical and the small nuggets are lying on the bottom. There it was, two little nuggets lying on the bottom!

Well it sure ain't no Bonanza, but it feels pretty good and I'll take success anyway it comes.

Hey Whoa Up there Tom, what's with this big red handkerchief and where does it fit in? Well I could say, it's for crying the blues if there ain't no gold, but really it's to spread on the ground beneath or within the crevice where things are too tight to insert a scoop or spoon to collect the contents and dump or wash in your gold pan. Most of the contents of any crevice can be collected or at the least thoroughly inspected for gold pieces using this process.

Sometimes a rolled up wet handkerchief will really do the trick as it sticks to the almost vertical sidewalls and will capture the material which frequently contains flakes and fine gold. The more crevicing you do, the more innovative you become at "Gittin the Gold."

Well, now you say, this seems to take quite a while and your chance for success is very limited. First, let me tell you, that any time spent in the outdoors and on a river is a success of value for anyone in today's world. I've never come home skunked for gold on any crevice excursion, but I can't say the same for some of my other ventures. As for quantity of gold, Tom and Louise, of Elk Grove, Ca. could show you a two ounce vial almost full of small nuggets and pieces they found during their two week vacation. I think they've given up any other type of gold mining in preference to crevicing.

As for myself, I can spend a whole day looking in these small hiding places and never tire of the effort. Success?? My last day out, netted two clinkers, with very little work and a great deal of pleasure involved.

Normally my gold mining efforts involve WISS BANGR High Banking or Dredging, but occasionally I get away for a little crevicing.

I still have about an ounce of what I have found over the past few years using this technique. The largest crevice find (nugget) weighs in at only 1 1/2 dwt., but then again, I'm not finished looking!!!

Chapter 27

Frannie Packer Gold

Most gold mining exploits begin with some kind of a challenge and this one was no exception. First, though I have to say that it don't matter weather you're panning, sluicing, high banking, WISS BANG'ing, dredging, drywashing or jumping someone else's claim, if there ain't no gold in the dirt, you ain't gonna git none!! So the most important part of gold mining is prospecting and locating the gold, after that you determine what process you need to use for gittin the gold. This is just such a story.

I was prospecting down river of Slaughter House Gulch. The immediate area surrounding this Gulch is only accessible by means of ancient and difficult mountain trails that the old timer's used and maintained for access to the working area. Most of these had long since become dangerously unusable and this condition prompted me to seek some means of getting into, or across river past this area.

I found it was necessary to get into the mouth of Slaughterhouse Creek to cross, as the rapids down river were too rough for a safe crossing. It was during my first exploration down river of this area, that I came upon a fairly rich deposit located on the river bank that covered about 600 ft of shore line. Well,this was pretty exciting and the average gold miner would have immediately set up there and started an operation. I would have also,except that I didn't feel up to carrying a ton of dredging, high banking and rock moving equipment into the area across a steep and dangerous river bluff.

Additionally, my wife Fran was otherwise engaged picking up gold in another "glory hole" at that time. So off and on for the following few years, I consoled myself with other mining, but kept a strong weather eye on a box shaped car sized rock upriver that I knew held some good prospects. Occasionally, I would move upriver for a quick look see with a gold pan, but; not with enough equipment to make a real mining effort.

Coming into the 05 mining season, I wanted to make a strong effort to mine around that box car rock, but once again too many things worked against me to move my equipment. I did resolve, however to shoot for the 06 season. Towards this effort, I dedicated many hours to hand digging a trail along the river bluff from Fountain of Youth Creek to Slaughterhouse.

As a side note of interest: Some folks tell a different story, but Fountain of Youth Creek (my version); was so named because you take one drink from this creek and you may not get a day older. Really though, the water is pure bed rock generated and of excellent quality.

This river bluff was a killer to navigate, nothing but jagged rock spires,shear walls and fairly deep crevices. Some areas were blocked by huge boulders and rock slides. Things didn't look too hope full and any headway, I made, was very minimal. I did kinda scope out the best method to build a foot path with hopes that I could figure out how to get around most of the obstacles. Going up the mountain, over and around was just outta the question. That over the mountain business was a much too difficult and long trip with heavy loads.

That winter during my stay at my home base, I started working on a plan that would aid me in getting upriver the coming summer. My efforts involved locating some kind of an economical powered unit that would carry 150 pound load of equipment over a narrow footpath type trail. There wasn't such an animal being built by anyone that would do this kind of a job. All of the research was of no avail, no one built a machine that could navigate a crooked and rough 14 inch footpath trail with any kind of a pay load. I had to build my own.

When inspiration comes to the aid of a person with the non-working (lazy) inclinations that I have; it's easy to come up with a solution. According to my thinking, a small cultivator, properly modified might serve the purpose. It appeared to me that low powerful gearing and centrifugal clutch operation were two of the basic requirements and the small cultivators had both. I went looking for a candidate. My wife, Fran is more than well aware that some of my contraptions don't always work so good, but; I scoped this effort out pretty thoroughly ahead of time and she didn't look too hard at my $350 investment. I really had no need for a cultivator and I wasn't about to go gardening if this thing didn't work,but if it saved me from six weeks of therapy due to shoulder stress injuries; it was gonna be well worth it.

I bought a pair of solid wheels, took the tines off the cultivator, removed the flat axel drive hubs and mounted them on the wheels and now I had a two wheeled power driven "something or other" that would run at 4 mph. I added some 3 1/2 ft. rails and one 14 in. rear wheel off my defunct lawn mower, installed an iron milk crate on the forks of this contraption and "VIOLA"!! "The Frannie Packer" Now!! I'm ready to go upriver gold mining.

"The Frannie Packer'

Well, 06 sure didn't look too good, the water was high and promised to stay that way on up into August. In early May, I started working on Rattler's Crossing so named because in 05 the little portion that I had finished produced the death of a few rattlesnakes. They really seemed to like that Fountain of Youth Draw and that was at the beginning of my trail. Mr. Bob Yawn, good friend and associate gold miner with 12 or size 14 boots tromped right on one just in front of me. The little beggar was only 15 to 18 inches long himself and Bob's big foot sorta squished him a little and I quickly dispatched him with a handy rock.

Sometime later in May grandson "Cowboy Tils" came to spend a couple of weeks,so I put him to work helping me haul dirt for the trail. He did a great job and in fairly short order, we completed a bad slide area. I was able to fill in a few big cuts, move some boulders and with Fran's help finish the trail all the way to Slaughter House Creek. Now, the Frannie Packer (so named to replace Fran, whom had been serving as my mule for packing things upriver) came into it's own.

This was the perfect transporter to move the winch, power motor and all the heavy rock moving equipments. As a side note; this little transporter also served to haul lots of food and drink for the many pot lucks in camp during the summer.

The 06 season came on with a bang, a completed Frannie Packer Trail and access to the Box Rock Glory Hole across river. We, Fran and I had already taken a couple of ounces of gold from this area and the word was out, head up river to the Box Rock Bonanza!!

This kind of news in 2007 brought on a super new gold rush with "Marshal"Neil Dixon leading the way to "Dixon Diggins' along the Frannie Packer Trail and much more Box Car Gold. A host of friends and fellow miners joined in this Quest for Gold and the good times. A Bee Hive of activity thru August and September of 07 saw high banking, sluicing, panning, dredging and metal detecting success at the Dixon Diggin Site.

Everyone now, is looking forward to many more great fun year's when the trail is open, the food is wonderful, the gold is great and the "Frannie Packer" rolls again!!!

At least 40 trips up river, and two tons of equipment hauled; this little transporter sure earned it's keep. Everybody wants one, but I'm not the guy to ask; remember I'm too lazy to work and build these things! After all, a feller has a reputation to uphold.

Chapter 28
Bottle of Gold Poker

History is full of the gambling free spirit of the gold miner and his tendency to "Strike it Rich" then gamble it all away in a card game. A common explanation for coming home broke is "I lost all my gold in a poker game!" For whatever reason, that explanation seems to be acceptable and understandable. A gold mining "Bum" is one thing but a gambling gold miner is something quite different.

One of the most satisfactory and enjoyable operations that I was ever a part of resulted in the development of a family type game called "Bottle of Gold Poker" on Monday Night at the Roaring Camp Trading Post. This event began with an instruction class in the use of the Wiss Bangr High Bank gold mining method of washing bed rock and collecting the material for classification in a sluice box. The operation and techniques employed are highly successful in producing free gold with minimal work involved while maintaining an environment protection process through the use of recycled water and a settling pond. I sorta fell into the development of this instruction class and it proved very successful, so much so; that every Monday morning I wound up with a (sometimes not so small), bottle of gold. Since I employed all of our visitors in getting the gold, I felt they should have it.

So, going back to yesteryear: like 1850. When a company of miners numbering three or more, worked all week mining; and their take after expenses was small, they played poker for the gold and one of the company would then have enough to go to town for a week end blast! This Bottle of Gold Poker Game had it's beginning modeled after that process however; the game as we play it today, is a family game of chance; not a gamble. The following story is an event involving one of these Monday operations.

We were going to work bedrock on the old ancient Cleveland Channel near the site of Caesar's Mining Claim of 1851. Caesar's Claim is now part of the Roaring Camp Mine Resort and as such provides a mining prospecting site for vacationers. The Cleveland Channel is a pre-ice age river now buried under megatons of mountain material for millions of years. A large number of mining towns within the Mother Lode area, sprang up as a result of this gold bearing channel surfacing near by. Geologists informed me that much of the material in the adjoining Mokelumne River is a part of that old channel and dates back in time to pre development of the Rockies. Now I ain't much of a hand at this geologist stuff but,I did inform Mr. Geologist Dave of Bakersfield whom was quite interested in a hole I was digging, that although I appreciated his love for these fantastic formations etc. there weren't no gold in them. He says, "But Tom, these are amazing and wonderful to study in their natural state!" I said," Yeah Dave I reckon that's all astounding but, there still ain't no gold in there!" I'm not sure he was aware that I was stringing him along.

Well, the history of this ol'channel with it's multitude of tunnels,is quite interesting but, we were here for gold mining. I explained to all that these ol'tunnels of the 1800's were not safe for exploring and in answer to a question as to their depth into the mountain? I said, "I'm really not sure, but during one of my explorations before I developed good sense, it took me ten seconds at a break neck run to get back outside when a rock fell!" "So, I can tell you for sure it's more than ten seconds deep at a dead run!"

The face of the bluff is a lengthy exposure of the bottom of the old channel and it is gold bearing. This exposed face has been mined and produced gold since 1850. I had taken a few pans from various bed rock areas along the bluff and selected a potential producing area where the mineralization showed to be very heavy. Dead zones in todays rivers show a lot of this similar mineralization and consequent gold drop areas. You know seven million years don't really make much difference to gold, it still acts the same.

We sat up the equipment and proceeded to wash and clean the bedrock. There were thirty four workers and, all put in their mining time on this potential producer. We gave each worker at least a five minute introduction to Wiss Banging and then told them they were fired. Three times the hose blew off the suction nozzle drenching everyone and filling the hole with debris and muddy water. This is a common occurrence when the hose and nozzle are handled improperly and frequently leads to a lot of friendly kidding between all that are soaked. The real reason for Monday morning Wiss Bangin is the socializing and friendly ice breaking that takes place between strangers.There was a lot of debris and mud all around. It really didn't appear that we were going to have a good show for our efforts. I asked Joanne LaMott (experienced gold miner) of Texas to assist me in doing the final cleanout and inspection of bedrock.

We cleaned out the fallen over burden, rocks and debris and even though we were using recycled water, the bed rock was clean enough to eat off of. Then—THERE IT WAS!!! gold in the bed rock!! I picked up a nice piece and then tried prying more out of the bed rock. Joanne was shaking so bad she couldn't pick up the gold. She said, "The hair was standing up on my arms and I couldn't get my breath; I've never had such a feeling before!" Said that she and husband Paul had been mining 21 years and never seen anything like it! A fantastic experience of a lifetime and a bottle of gold like rarely seen before; on a Monday exercise!

Did they all show for the Poker Game? You better believe they did, all thirty four itching for that Bottle of gold. Five card draw poker is the game, three hands each, with top hand winning the table,same game for semi finals and finally the same game to win the gold. Going all the way to the final table was Joanne and three others. Joanne was holding two fingers on the gold and standing first, when I asked for the last call. Young Miss Taylor Larson stood between Joanne and the gold. I said "It's your call Miss Taylor, How many cards do you want?" She said "Three Tom, and I want'em from the Top of the Deck!"(Whoa!! She was kidding, maybe?). Well, to make it short ;with no help from the other thirty three players and winning all the gold was this pretty Miss "Taylor Larson" of San Diego, Ca. Couldn't happen to a nicer young lady! "Them Pretty Girls Always Get the Gold"

"The winning hand with the bottle of Gold QQQKJ"

Chapter 29
Big Rocks, Big Gold, Little Rocks, Little Gold

Rocks are a fascinating subject and a large number of folks across the United States are members of Rock Hound Clubs. Their hobby and enjoyment involves the finding, collection, and development of beautiful art work from the virtual millions of types and formations. You know I appreciate and enjoy these folks because a number of them are friends and even relatives. I'm also a rock collector of sorts; but, I prefer those allegedly referred to by our early Native American Indians in the statement, "Crazy White Man Looks for Yellow Rocks" yellow soft metal ones known as Gold! Some folks call them nuggets.

Where do rocks fit into this picture? Well if you took the advice of a good friend of mine, he says quote, "A good day mining is when your partner doesn't hit you in the head with a rock!"

As a placer gold miner, rocks are a major part of every gold mining operation. They are always in the way! I'll admit that I've found free gold laying on bedrock just waiting for some ol'gold miner (like myself), to luckily stumble over and pick it up with their fingers. This,however is a very rare occasion,because in most instances gold is covered or buried on bed rock and frequently involves the movement of large volumes of over burden to access. Most generally rocks are included in this covering and in most cases huge boulders are involved which hide the gold. Maybe our friend and local Gold Mining Confucius, "Mr. Kanji Gotori" is right when he states "Big Rocks, Big Gold; Little Rocks, Little Gold."

Now there are scientific reasons for the development of rocks and the classification of same. There are also formulas available to enable you to determine the approximate weight of rocks both dry and in the water. In my case, I'm not really taken with the development of rocks or the formula's. There are only two classifications that I'm concerned with. If my wife Fran can move them when she's dredging, then they are (in my mind considered rocks), if she can't, then obviously they're boulders and I have to get involved. This, of course makes the job a whole lot easier, because I don't have to get into all that scientific study stuff to determine and characterize the boulder or generate an engineering logistic plan before I slap a choker cable around it and drag it out of her dredge hole. There is another major personal advantage for this classification. (Fran moves all the rocks!)

Boulders come in many shapes and sizes. the most common characteristic of all being they're big and heavy. Looking back at the 49'ers, many written accounts from that Era indicates that when the miner came upon a large boulder blocking his way; he got a sledge hammer and began beating it into small pieces. (you know I'll bet the Warden of Yuma prison that introduced the rock pile for inmates was an ex Gold Miner). There is physical evidence of this boulder distruction found in the artifacts collected near mining encampments, such as the remains of the sledge hammers and chisels that have definitely seen better years.

Further study reveals,that when a major gold strike involved a boulder field; a company was formed with manpower available to operate rope block and tackle and huge "L" type leverage arms for lifting and moving the boulders.

Large areas in the river were cleaned in this manner and some can still be seen to this day. Many areas in gold bearing rivers show huge rock piles that didn't get there through natural river action.

Well, I guess it's time to pass on history, quit beating about the bush and get into moving some rocks. Most of you gold miners know that there are no roads along the river banks except in those areas that have been hit a thousand times, so in order to give yourself the best opportunity to find "Big Yeller' you're gonna have to go where the roads ain't. That means,of course you're gonna have to carry in the necessary equipment to do gold mining and to move boulders. (Remember the wife moves the rocks) Yeah! Right!! So, lets select the best and lightest weight boulder moving equipment to get the job done.

The Grip Hoist is a hand operated cable puller type of winch far exceeding the capability of Come-a-longs, Chain falls, and Hand Winches. When used properly in conjunction with pulleys (sometimes referred to as chives or blocks), one person can move tremendous size boulders.

One of the most common in universal use is a 4000 lb rated Grip Hoist. This unit weights 60 lb with the 60 foot cable. I usually carry the puller separate from the cable to lighten my load. There is an additional couple of hundred pounds of cable, shackle's, and separator plates that are necessary for pulling boulders. Aside from the Dredge with motor etc. all of this paraphernalia has to be hand transported to the working area, usually a good ways from your vehicle and consequently requiring many trips.

A gold miner accepts this as part of the necessary "pleasure?" associated with "gittin the gold." Man this recreational activity that supposed to be good for you sure involves a lot of work!

Hey! Whoa up there Tom! How do I find this working area to start my dredging? Well, a possible gold drop zone in the river is on the inside bend, an abrupt widening of the river channel, downstream of large bedrock outcrops, large boulders, or in a lot of cases the presence of hardware underwater or on the shoreline. These are all good indicators or potential gold locations. The very best,however is where other folks have had good finds before you. I don't believe that ol'saying that gold is where you find it. That's kinda trusting too much and a lotta hard work to go by blind luck. I believe this old "Tom Leftwich" truism, that gold can be found close to where it's been found before. Nobody ever gets all the gold and there just might be a big boulder close by a producing location, that has never been moved.

Are you lucky? The first thing you run into as you start your dredging operation is a big ol'boulder. Is that bad? Some folks would think so, but; here is where your education begins. This big ol'boy has never been moved before, at least in my humble experience and opinion. The exposed area that I can see shows low water ring discoloration generated during recent drought years, and these are parallel to the present water level. This indicates no recent movement of the boulder and the absence of miniralization marks or exposed common hardware rust spotting contributes to this belief.

Fran says I'll clean around it and you figure out how to move this "little" pebble. So-o-o-o, Let's just say by way of comparison that this little granite baby is 8ft x 8ft and generally round. That equates to about 40 tons or 80 thousand pounds minus up to a little less than half of that weight because it's under water. Oh Joy! It only weighs 25 tons. Why I'll roll this little Junior pebble out with my pry bar. WRONG!!! Get real, we need to project this baby.

This means I try to figure out which way to pull and find me a dead man (a boulder big enough to take the load of this boulder) and another to which I can attach the Grip Hoist. I never put the Grip Hoist in line with the load or dead man as this could be suicidal if a cable breaks. Then I have to pull everything else out of the way to get to Junior and some of these smaller boulders can easily be of multiple tons in weight and require individual attention for clearing the area.

Once cleared, I know from experience that the drag weight of Junior, (friction against bed rock cannot be determined but it's gonna be bad) so I'm gonna use my 6 block configuration to try and get the job done. This means that I can maybe put 8 lines to the load and develop 32000 pounds of pull against this junior pebble. If I need more,then I block the load line with a heavy duty pulley and ¾ in. cable effectively doubling my pulling power. This increase requires the addition of another dead man for the ¾ in. cable Things are really getting serious now and dangerous at this point because you're generating 64000 pounds of pull and ½ in. cables pop like shoestrings under these conditions, According to my experience, barring real hang ups against boulders that I haven't and should have moved, Junior should at the very least tip over allowing Fran dredging access to where it's bottom used to be. With a little luck,there might be some gold underneath.

"Here's just such a big boulder turned over by good friend Don Miller, and the bottom is now looking upriver!"

So here we go with a multiple pulley/Grip Hoist rigging and set up that we use to manually pull this boulder over in order to dredge out the bed rock underneath. We never undermine a boulder. The chances of it coming over on top of you is not a pleasant thought. Many accounts of coyote hole cave-ins that killed miners, fit this very category; when miners digging on the banks and bluffs tunneled in under big boulders. Undermining is definitely a no-no.

Well, we have this big boy laying over and we insure that it's not going to come back on us. If I feel there is any possibility of this happening, I'll drag it completely out of the way or chock it with other boulders.

Now all of this effort requires time(sometimes weeks) and we still haven't seen any gold. This is correct but, remember the gold isn't going any where and there very well may not be any, but; we now have a safe opportunity to have a good look. It's all in a day's work for the gold miner and if this was a dry effort, well, I tell Fran, "Hey! There's another good looking bunch of boulders just up stream a little ways that have all the earmarks of a "Real Glory Hole!" Then I duck the rock she throws!!

Chapter 30

Reading the River for Gold Drop Zones

Most major libraries and all Gold Mining Magazines or Stores are a great source of information on all aspects of gold mining. Always at some point, will be included, the nature of gold movement and potential gold drop zones in any river channel. There is a wealth of excellent knowledge available on the subject of finding gold. I for one, want to thank all those folks for taking the time to put their knowledge in written form for beginners like myself, because; I started as a greenhorn in 1980. Fortunately, I've had the good luck of working with great gold miners and prospectors over the last 25 years and some of these old timers couldn't read or write a lick but, their personal stories and experiences give a lot of credence and creditability to a great deal of what has been written. This chapter covers some of those things of value, that haven't been written.

Many gold miners spend a valued amount of time looking at the river trying to determine a possible location for gold. I sometimes do the very same thing but, I've learned to try and imagine a river action that will move and deposit the gold. As an example; looking up river in a number of locations, observe that the water is turning to the left and if you followed the advice given in most books on gold mining, you'd naturally think that gold deposits were on the inside bend of the waterway. This would be the norm and in a lot of instances you would be correct, however in some instances you're wrong. You're looking at summer water which is running about 10 cubic feet per second, or less; past any given point in the river bed. Now take another look and imagine flood stage at 3000 cu.ft.per sec. That puts the water up on the side walls and notice the side walls indicate that the river turns to the right in this particular area. Gold will move during flood stage, especially the larger values. So what does that tell you? Let's go mining on the right hand side! That's where the better gold will settle. This summer water won't hardly move fine gold. This is a typical mistake made by a large number of green horn gold miners that need a little help in determining a location for mining. Finding gold is hard enough! Let's at least start our operation in a favorable location. Once again I emphasize, always give yourself the best opportunity to win.

I follow my own advice and try to use a very objective eye when looking at the river. Always try to imagine flood stage and where do I look for a low pressure area that occurs at that time. Many of my friends and associates have given me a funny look when I said I'd dig over there when it appears to be the wrong side of the river. By the way, I never tell anyone where to dig, I generally recommend based on experience and judgment. To tell is to accept blame for no show and frequently that no show is due to lack of effort,not lack of gold.

Then there's another tough call in an area where the river runs slap dab head on into a sheer wall then turns say to the left. What do you make of that,Mr. Gold Miner Tom? Well, put yourself theoretically in a large pipe filling with water and what happens when it fills up, the filling unit reflects a back pressure which effectively stops the incoming water creating a low pressure area. In a river bed that feeds directly into a wall the same back pressure will occur, a low pressure point will develop and that's where the gold will drop out. In just such an area, guess where the gold miners have been pulling out multiple ounces of gold the last ten or more years. You guessed it; right out here upriver where the reflected pressure from the bluff butts up against the onrushing flood waters creating a veritable dead pressure zone. "Hello Gold!"

I'll attempt to describe this river flow based upon my personnel experience in just such a location. Let's take a look at another area that I'm very familiar with, which has the same general topography. The river in this instance hits another bluff which reflects back pressure up say, to where I'm standing, This particular area some what upriver of the lower China Camp Falls, always produces gold for any dredger willing to give it a try. It's location and the difficulty involved sorta discourages most folks, but to this day no one has taken it down to bedrock. John and Judy of Scottsdale Az. picked up ½ oz here basically from the shallow overburden, and I have personally taken 3 dwt from the side wall just crevicing. Does that interest anyone?

One more area,I want to cover because it represents another situation. This area illustrates a choke location where the river is necked down and then drops into a wide flat pool. This area is known as Jump Rock Crossing and the pools immediately down river always produce gold and a healthy vacation for dredgers. All the gold that comes down the north fork has to pass through Jump Rock and there immediately below is a low pressure drop zone. This would seem to be a great attraction on it's own but, looking a little further down river we meet the confluence of the South Fork and the North Fork. Bang!! The North runs slap dab into the South at right angles, both in flood stage and here comes that reflected back pressure generating a beautiful low pressure drop zone in all those nice pools below Jump Rock. Yes many dredgers have done well in this area, but most had little or no rock moving equipment. So-o-o, Does that tell you anything?

Moving right along, we're going to deal with another peculiarity I discovered during my earlier gold mining days. There was an area down river of an abutment on an Island that captured my attention because it was covered with round rocks. These were all in that 5 to 10 inch size and I asked the local Gold Mining Sage about this strange deposit. There were literally hundreds of them covering the shore line for a significant 100 feet or more. He related to me the fact that he

believed the ol'timer's (49er's) had classified them and placed them there. Well, there's nothing in the history books to indicate the miner's were of a mind to waste their time doing something like that; so I did some exploratory work in that area and formed my own conclusions.

This is just my own opinion based on what I've seen and experienced on this river covering the last 25 years. This deposit was mostly of round or rounded rocks and illustrated a deposit at the upriver end of smaller cobbles with the same physical characteristics. My investigation found some hardware mingled in with the rocks, which leads one to believe that this is a low pressure drop zone. Round rocks offer the most weight versus low opposition to water flow; in other words round rocks act like gold nuggets versus flakes of gold, they tend to drop out earlier while flakes with more surface area(like flat rocks) are carried on. I have not had the time or opportunity to verify gold deposited on that shore line but, I've got a full two ounce bottle, I'm willing to bet on it.

Well, I've covered a few of the pointers I wanted to pass along as I have seen evidence of it's truthfulness. I feel that most miner's know of similar area's on many other rivers and I hope that these descriptions are beneficial—"Did I forget anything??—The exact locations where these spots are located??" Oh Yeah! Well, Sometimes, I tend to omit a few important details on purpose. Remember!—I'm a Gold Miner too!! Good Luck!

Chapter 31

It's the Gold I'm After, Easy Ain't What I'm Here For

Fran, my wife said: "Have you finally lost it completely? You think there's gold up there? Are you outta your mind?" We were working below the South Fork Falls on the Mokelumne river and I was eyeballing a large outcrop of bed rock that protruded up from the side wall of the river bed. The top of this formation was a good twenty feet above water even at high flood stage and the opportunity for gold to be deposited anywhere up there was sheer fantasy. What she didn't know was, while crevicing; I had found a nice little one pennyweight nugget up on top of this outcrop the previous year, and my examination of the whole area had shown a number of pockets that had been formed by decaying quartz deposits in the base bedrock.

The problem at that time was, the pockets were too large for just crevicing and I really wasn't up to bucketing all the material to the river for a sluice operation. I explained to her that my over riding tendency to look for less work prevailed last year and so now here we are with a fresh opportunity to explore those high pockets. Her Question? "I suppose you expect me to tote those buckets; what with your bad shoulder and all?"

"Well No, I was planning a high bedrock Wiss Bangr operation that would allow us to wash and clean out those pockets." I admitted that the almost vertical rock formation would be a real challenge and it would likely be a slow operation. I explained that the high steep walls had kept most folks away from these pockets I was interested in, and for that reason they offered a better than fifty-fifty chance of being producers. The fact that they were high outta water and couldn't possibly see any material flow except during a thousand year flood was plenty of reason for most folks to pass them by for more likely locations. That nugget that I found though; got up there some how; maybe when that area was the bottom of the river or flood stage was much, much higher; and a number of others might of stopped there too. Since it was impossible for us to cross the river, at this time; with it running about three feet above normal, why we might just as well get a little mining in on this Little Rock Mountain.

Fran was not impressed, I think she still preferred that "Are You outta your mind?" opinion!

The downstream side of this large outcrop consisted of a huge trough that basined toward the bank and sloped gently towards the river. Mining friends Robert and Neil of Los Angeles, had previously worked this area finding a little gold in the basin area and Bill with wife Cheryl of San Diego; had enjoyed a good vacation mining the trough to the river. They had a great time and each day brought a good return in gold. I figured that this area,at the river level had been hit by the ol'timer's of 49 and hit again during the depression of 29 but, knowing that gold was still being found there; gave me encouragement. Fran and I could have a shot at this Little Rock Mountain with a halfway chance of getting some gold!

The basin at the foot of the steep wall was a perfect settling pond for our tailings and the trough would carry our excess water to the river. Hopefully we could develop sticky feet to help us stay up on the pinnacle where we would be working. Those sharp spikes and formations in some of these bedrock pockets had kept many a miner from the type of exploration that I had in mind and I always get more excited as we get into an operation. This place was definitely a challenge and we just had to figure out how to beat the bad odds and win.

Our operation was full of difficulties right from the git go. There was no trail by which equipment could be moved in by two people sharing the load. Transporting a five horse motor into the area was kinda outta the question. We elected to use a small back pack 2 1/ 2 HP Wiss Bangr and although it would be a little slower with vertical pressure reduced due to the high lift required; we could get the job done. Another major concern was the nature of the river. It was at flood stage and could change hourly because the South Fork was a wild river(uncontrolled); very much subject to weather changes in the high country. The safe location of our motor on the river was something we always considered and we always moved it and all of our equipment way up on shore for overnight. Lessons learned the hard way (river rising nine feet overnight and taking your equipment downstream) tend to stay with you.

The location of the sluice box was a real concern, because the basin could catch all of our tailings and water; but we still had to have a location that would permit level set up with the proper tilt angle for proper material flow. Sometimes,this is even difficult on the shore line,much less twenty feet in the air on top of a pinnacle of bed rock. I guess we got lucky or ol'Mama Mockaloom (Mother Nature of this river) took pity on us. Between the use of rocks and natural formations, we were able to get things in some semblance of working order and turned on the water. Finding footing to start an operation was kind of precarious, but pretty soon muddy red water was running through the sluice box; and now it was time to get to "Looking for Gold"!!

Our first operation was a near disaster when I shut down the motor to realign things. (I should have known better). The high lift of water to our sluice box and suction nozzle caused a loaded back pressure on the foot valve attached to the water pump when the motor quit running. This back pressure blew the foot valve off the end of the suction hose and "good bye" in the flooded river. This was not the first time that I've had this experience and normally I'll attach a retaining cord or wire to prevent losing the foot valve. Dumb, Dumb, Dumb!! All of my other motor's have this attachment and I always remind all of the other miner's to use a safety retaining cord when they set up. Enough said! We lost a day's operation and a perfectly good foot valve.

So far this operation was developing all the ear mark's of a Snake Bitten Project. Sometimes you might think that bad luck and bad judgment has got hold of you. Fortunately, all gold miner's are optimistic and tomorrow's the day for that big nugget find! That gold won't go anywhere and we'll be back at it early in the morning just as soon as I can find another foot valve. (and a retaining cord).

Why do I get involved in these near to impossible projects? Think about it, all the easy ones have already been done three or four times ahead of me and I always remind myself, "It's the Gold I'm after; and Easy ain't what I'm here for."

The operation had started off with mistakes and misfortune, but it turned into a veritable adventure of success in learning how to High Bank (Wiss Bang) on a rock pinnacle and clean out those never touched knurly pockets in bedrock full of splinters and spires of sharp formations that have protected these pockets from previous cleaning. We initially washed and cleaned out the pocket and then employed the use of the air hammer to turn it into a smooth bowl for cleaning. All of the adjacent pockets were treated the same and within a relative short three days our operation was completed. Was our project a success? Well we didn't exactly win the Lotto, but I sure don't complain about a nice bunch of fines and a few gold nuggets! How does a little over 8 pennyweight "Grab you?"

"Climbing out of a 'CLEAN' Five Foot Grind Hole"

Chapter 32

The Big Gold Strike at the Fence Line

As related to me on numerous occasions (sometimes after a few drinks) by Leo Wendt, Bob Scruggs, Roy Scruggs, Jack Neal and a large number of collaborating family members and Roaring Camp miners.

My association with the Roaring Camp Resort began in 1981 and has continued till the present time. Many gold stories were told to me by the Sages of Roaring Camp and long time visitors that added insight and detail to a glowing history of the adventures associated with some of what we now call the "Old Timers." Sadly most are history and due to their lack of reading and writing ability their individual adventures in gold mining are forever lost. Fortunately, when I came to camp, some of these old timer's of more recent origin were still kicking and I was able to record their wandering stories in as much detail as they were capable of remembering. One story stood out and I interviewed all of the participants and every associate to which the story had been told. I recorded as much detail as I could get,because I knew that someday; I intended to put it all together. Only those coinciding recounts of events by the participants and their associates are included to insure accuracy.

The story that captured my attention involved participants that were to become great friends of mine and involved a great gold discovery made about 4 miles down river of the Roaring Camp

Mining Claim. This is their story and my tale would never be complete without a recount of the biggest gold strike ever made on the Mokelumne River since the mid-1800's.

In the early sixties the brothers Bob(aka Buddy) and Roy Scruggs and brother-in-law Jack(aka Jackson) Neal came to the Mokelumne Canyon and what is now Roaring Camp Resort. The Camp at that time was a series of deep cuts into the meadow area where one had to walk down into each and wade through the muck and swill water to go up or down stream. These deep gullies and cuts had been made by the Clinton Bar Mining Co. cleaning bedrock coming off river. Like all mining operations,in those days,everything had been left just as it was in 1942 when the Government shut down mining for the WWll effort. The camp as we see it today was a complete mess that you couldn't give away. The private property existing at that time (a patented mining claim) was owned by the Elton Rodman family and he had given permission to this group to pass through his property for access to their mining area.

Bob and Roy were from Mississippi, which had; not the greatest of gold mining history but,working in the California gold area's, they had earned their degrees in gold mining through application; on the river. Roy, at that time had a family in the town of Plymouth and it was probably his knowledge of gold mining in the local area, or lack of it, plus gold fever, that inspired his partnership in this adventure. Jack Neal had married into the family and he had been raised a gold miner in the Arizona gold fields. It was only natural that he wanted a part of this family action. Bob (burning with gold mining desire and innovative inventor of the first underwater dredge used in northern California) was the promoter and it was his association with Matty (wealthy yacht shop owner) of Long Beach that grub staked this gold mining effort. One must say that Bob could tell a good tale(still does) and inspire anyone to feel that a Bonanza was within arms reach for just a little investment. Expectations were very high and so the adventure began.

At this time Elton Rodman of Pine Grove, was putting together what was to later become the Roaring Camp Mining Company and Resort. The famous Fence Line near the Ponderosa Bridge was to become the downstream border of Roaring Camp. Who owned it at that time is probably just a guess as possession was sometimes determined by who held the gun. It was up stream of this fence that the Scruggs Boys and Jack started their dredging operation. They had built a small boat with a Renault four cylinder engine that powered a large water pump which had a great versatility for powering them up and down river and in addition provided the necessary pressured water stream to operate the underwater dredge.

It was Bob's original underwater dredge that was used during this operation. High pressure water delivered at the front end provided the necessary pressure release required to generate a vacuum at the inlet, ingest the material and deliver it to the enclosed sluice box where it was classified by the enclosed riffle box. Stability and vertical position was maintained by adjusting ropes attached to floating inner tubes on the surface. This was the very earliest type of suction dredge used on Northern California rivers and was the brain child of Bob Scruggs. As a Note: Many of the mining techniques and equipments that we use today were developed in a similar manner without recognition or monetary reward afforded the originator.

Things did not go very well for the adventurers, They were struggling to make enough for gas money and "Money Bags" Matty was not too well pleased with his rewards for his investment. Also, I was told that Roy's wife was not too pleased with what was going on with their gold mining efforts and she sent demands that Roy come home. On one occasion, he did go home to try and put out the fires; but returned because, like Bob, gold fever can get to be an addiction. Jack Neal, in the meantime was practicing the art of making grape wine from the wild vines along the river and trying to help keep things going by bringing in wild game to add to the canned meat, sardines, and beans. Jack was a dead shot with the sling shot and at that time Quail, Dove and Robin breasts barbecued pretty good. It's understandable that home life, irate wives and family, for all three, could not totally be offset by gold fever.

Things were going downhill with the gold mining and their efforts had not produced more than a few ounces of fine gold, This return for days and days of dredging. They were,in all three opinions; hungry, dirty,poor, and tired of digging. Pork and beans, bacon and eating out of a can was definitely getting old. Bob had left to go to Long Beach and try to scare up some more money, and had told them to move their operation to another likely looking spot he'd picked out. Roy and Jack figured they'd quit this area at the end of the day and relocate the following morning. They were not looking forward to another move and the beginning of another possibly dry hole operation.

Jack had been taking a break while Roy was dredging and it was now time for him to take over. He grabbed his huka and weight belt, slipped underwater and headed for Roy's location. They had been cleaning out a crevice in bedrock where it appeared that a seam going at right angle to the river, had broken out at some period hundreds or maybe thousands of years ago. The trough was clean and Jack was following it towards Roy when he noticed a brown discoloration on the downriver sidewall. He took his rock pick and chopped at the brown wall and it broke away! There glaring back at him was GOLD! More than he'd ever seen in his life! He couldn't believe it! It had to be a dream! Maybe he'd drank too much of that wine! He almost drowned Roy trying to tell him about it!

They chopped out that hole,it was full of gold, a whole side wall channel laying full of all shapes and sizes of gold, nuggets the size of your thumb, beautiful pieces of quartz inlaid with gold and virtually thousands of flakes and small pieces. They didn't have a bottle that would hold it! On one occasion, Jack yelled at Roy to hand him down a bucket to put the gold in. This was the greatest day of their lives! It looked like a fortune and they had to get word out to some one. They had no form of communication, they walked out to get a message to Bob in Long Beach.

Bob came a running. He told them to meet him at the Greyhound station in Stockton with the gold and he'd take it to Matty. According to all accounts, Roy and Jack met Bob at the bus station later that week with 16 four ounce (Alkerselzer) bottles of gold. Also according to all accounts, they met on two other occasions at the bus station with similar bottles of gold. This relates to 48 bottles of gold. How much was in the bottles? Jack and Roy said they were as full as they could get them, but some had pretty good size pieces with quartz and on their last trip, they told Bob that the crevice had ended and they were running out of the big gold.

As an interrupt to this story, I asked Bob about his activities after he received the gold and he admitted that he skimmed a little off the top to finance expenses and a trip or two to Tahoe and Vegas. I also learned there was a good looking tall long black haired lady in Jackson that got a pretty good handful of that gold. Additionally, there was a pretty big party thrown in Jackson for the other two gold miners, although it was alleged to me that no transfer of any gold nuggets took place between said miners and ladies,at that time. It is a fact though, that a number of the evening ladies were sporting big gold nuggets around town.

Well the story goes on! Bob finally came back to the river to see for himself and yes the hole was clean. They worked together for about a week cleaning up and on one of their last evenings at the camp site, each had a small baby food jar which they claim had a little bit of their private stock obtained from individual single mining efforts. (Maybe 5 to 14 ounces—after all a four ounce baby food jar could hold almost twenty ounces of gold) According to Jack and Roy, they all drank a little (too much) that night and Bob, worrying that they would take his gold; he got up walked into the darkness and hid his bottle.(Note) To this day,after many,many trips down river(with metal detectors) to the camp site and many hours of searching by the group, that jar has never been found.

Matty of Long Beach, never having seen any gold before, got to thinking he was getting cheated after all this gold showed up and then nothing. He counseled with whom Roy said; was a witch doctor and came to Jackson to find his Grub staked crew and see why the gold quit coming.

He accused Jack and Roy of stealing and they, resenting the thief allegations; immediately shut down operations. Matty, reportably; hung around for days driving through Jackson with his Witch Doctor who had a divining rod he claimed pointed out the location of all of Matty's gold. It's downright amazing that sucker jumped to attention at every Bar, Brothel and Gambling house in town.(could he have been right!)

Following the false accusations and threats to cause 12 Gauge bodily harm to Mr. Matty and Witch Doctor by an irate Roy and Jack, an agreement was made to convene at Matty's home in Long Beach for a gold split as had been agreed to at the beginning of this gold mining adventure. Jack went to the split meeting, but Bob was in Florida and Roy was in too much trouble with his wife to attend. Matty was a ceramic doll collector and was very proud of his collection, On the agreed day of the split, he was having a houseful of doll admirers for a party and when only Jack showed up, he didn't want to take time to divvy. He did however ;have his gold out on display in a large cut glass bowl with his loveable doll collection. He claimed it to hold 36 pounds of gold and according to Roy, each were to get a tenth with Bob getting twenty percent. The gold was never seen again.

Matty was well liked at the local watering holes around Long Beach and regularly attended them. After he came into this nice bunch of gold, he was prone to carry nuggets with him and show 'em off to his drinking companions. Unfortunately, he also bragged about his cut glass bowl full and may have exaggerated it's weight. Two,not so nice brothers followed him home one night and worked him over with a tire iron while, presumably; stealing all of his gold. By the time news of his assault had reached Bob and the boys, Matty had slipped into a paralytic coma and died.

Bob went to Matty's home and yacht shop to try and get some of their share of the gold. He had knowledge that Matty kept the best show pieces in his shop safe. Zero is what he found, the safe was locked but, when opened there was no gold. Matty's associates claimed that they knew nothing (except possibly the safe combination!) His home was the same story no gold and it had not been found in the possession of his murderers. What happened to all that gold and how much was taken out of that hole down by the Fence Line? Does anyone really know or has that knowledge already followed Matty to the grave?

As for Bob, Roy and Jack, for many years; they went down the river with a continuing search for another Big Bonanza, but the ol'bones grow weary and the trail gets dimmer. No one though can ever take away the fantastic memories of that day when these boys stood on the mountain and crowed. They had reached the pinnacle that all gold miners aspire to, that once in a lifetime unbelievable feeling of wondrous success and discovery that comes to so very few.

The victory and success in this story reassures all of us, that the possibility still exists for that great discovery for every gold miner that opens a crevice or moves a boulder or gets that strong beep-beep in his metal detector earphones. It's still the thrill of the hunt,and the heart stopping excitement that comes with every gold find that keeps us ol'miners going. The pleasure will always be there with the challenge and the quest won't be over till we are. Good Luck!!

Chapter 33

The PIGMI (Piggin for Gold)

Well now what in the "H" is a PIGMI? It's another one of those concoctions that is bred from the distorted mind of a gold miner with a bad case of the lazy's. (I know there's no proper spelling for this condition, but you gotta remember, this is a new git outta work machine!).

What does it stand for? Pneumatically Improvised Gold Mining Implement (PIGMI) A tool that employs air pressure to get the job done. The job is crevicing those hard bed rock indentations and fractures that capture the gold on all rivers, new or old,,wet or dry. Chapter 26 deals with hand digging crevices and they do produce a world of pleasure and enjoyment for lots of folks that don't mind a little physical effort. As usual however, I gotta find an easier way to do things, s-o-o-o here's the machine I put together just for the "Crevicer."

Along about July of 85 (that's "nineteen eighty five," by the way). I developed an idea that maybe if I had an air hammer down river, I could open up a lot of cracks and fissures that had never seen daylight. At that time,of course; all I had was a Wiss Bangr and it didn't lend itself to running an air compressor in conjunction with a high volume water pump, plus the added weight alone was a show stopper.

Earlier on, I think I indicated, that inspiration is a major tool of the lazy afflicted and I gave a lotta thought to light weight and how to get the gold. I finally determined that maybe I could vacuum up the gold if I could open up the crevice. My research was negative all the way around as a vacuum would pick up dust and small gold, but those heavy little pieces would stay in the crevice. Additionally, I still had no air hammer.

Fortunately, I ran into a "Blow Hard" friend of mine that knew a little about the Berneulli Principal involving venturie generation of vacuum through the use of high volume air pressure. This little bit of information pointed me in the direction of using an air compressor to generate a vacuum while providing the necessary air power for an air hammer. This idea took hold and I started putting together an idea and combination of hardware to build the first PIGMI (it didn't have a name at that time). The suction head became a modified sprayer nozzle with a kitchen sink spout attached to the suction end.

My first operational test of this air powered super sucker head came close to disaster, I had borrowed the use of a friend's shop model compressor and it could have been operating at 150psi. I attached a 1 inch delivery hose to the outlet of the sucker head and kinda laid it on table nearby. His shop was kinda messy so I looked around for something to suck up to test this new "whatever." I saw a ¾ inch bolt laying on the floor,so I triggered the air and Boom! That bolt came off that floor zip! And exploded out the end of that tube at Mock Do Do, ricocheting off the walls of the shop and almost hitting us both. Jim-mi-neey!! Was I ever impressed! Boy! I was thinking, I could suck rocks up on one side of the river and fire them across. What Fun!!

Man, did that thing ever work! It was now time to get a design put together and build a unit to do the job. That winter I invested in a new small 1 1/2 HP motor and mounted it with an automobile rotary compressor on a back pack frame to transport down river. I tested it at home (breaking up some rock work for which I didn't get accolades from Fran). It worked Great!! I am now ready to carry it down river where fortune awaits me! I can't hardly wait for spring to get here.

Off to camp we go, and as soon as high water goes down,I'm packing this super sucker and air hammer to a bed rock glory hole before the rocks are dry. Fifteen minutes into my first operation and THERE IT IS!! A penny size 1 dwt piece of gold! Talk about happy, this thing is working great and I've already got the gold.

Then Clank, Clank, Clank!! The pulley and belt fly off my wonderful machine and smoke is billowing out of my compressor. My compressor had been running far longer than that home test and had frozen up. Later, much to my dismay; I found out that auto compressors carried lubricant within the Freon and I was using it for air without hence a freeze up! Well it's back to the drawing board!

I fought the battle and lost round one but, I went home and brought back my big shop compressor to try and finish the job. It almost finished me! That sucker weighed 175 lbs and I had to disassemble it to pack it down stream and reassemble on location. It worked great until I ran outta area to work. Now it's time for another pack it down stream effort that I definitely was not looking forward to. Fran came to my rescue she wanted to go dredging. Wonderful!! We went dredging and the PIGMI went on hold.

This idea hung around in my mind and the great results I'd had even for a short while. Crevicing is a fun thing to do and I continued to look for a solution that would provide portability and all the unique qualities that were necessary to make the PIGMI a usable machine. Fortunately, time and technology came to my rescue and I finally have a unit that is transportable and will provide the "crevicer" with a power tool that will get the job done. This new unit will pick up and transport nuggets up to 5 dwt. (larger ones you will just have to use your fingers!)

May of 2005, 6 year old Grandson "Cowboy" Tils assisted in cleaning crevices at the China Camp Caves "Piggin for Gold" using the newly developed PIGMI. His take was 3 pennyweight for a few hours work. Following that early use,I worked a high bluff area cleaning out one of those holes in exposed bed rock that are full of rock spires and splinters that make it impossible to get into for cleaning. That air hammer chopped out everything down to where the inside looked like a bowl and then ol' super sucker did the rest. A 1 ½ dwt nugget came outta there and I didn't bother to weigh the rest but, it was nice!

There are literally hundreds of high bed rock pockets out there waiting for me and the PIGMI. I'm itching to get down to a long exposure of bed rock just beyond where the river takes a sharp bend near what we call Lone Rock. There's year's of "Piggin for Gold" ahead just waitin for this ol' gold miner.

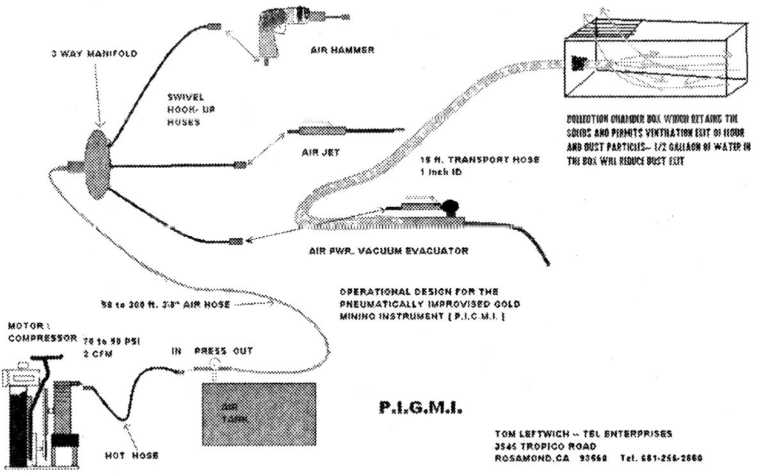

Operational Design of the PIGMI

Chapter 34

Where the Bucks Stopped

Gold mining adventures can begin in the most unusual fashion. This one began when the high water receding in May of 1999 disclosed that a tragedy had occurred somewhere upstream of Slaughterhouse Gulch on the north fork of the Mokelumne River. I was working at the Roaring Camp Mining Resort that summer, and it was during one of my journey's upstream that I discovered the results of this tragedy.

There in the middle of the river were the skulls and antlers of two buck deer that had locked horns in mortal combat, apparently fell in the river and drowned, their remains coming to rest just down river of our upper fence line. The story and tragic ending to their lives was spelled out in the still locked condition of their antlers. I was able to retrieve the skulls and antlers, cleaned them up, separated them, and later passed them on to my daughter, Loreea, whom used them for decorating her outdoorsman (husband Tim's) style living room. Now, one would think this would be the end of an interesting, though tragic event and maybe an unusual conversation topic but Wrong!! this was only the beginning.

"Horns of deer recovered from the river, the smaller set on the right were locked behind the eye guards of the larger buck—notice the wear marks in white"

My distorted (wife's opinion) and inquisitive mind has frequently led me into interesting situations, not always conducive to long longevity, though in this particular case, I don't think there was any risk involved. I surmised that if the bucks remains stopped in this wider river location, then other things like hardware and gold would also stop there. So, I took a very close look at the history of gold production from this area of the river and behold, I found a virtual wealth had been extracted from just downriver of the buckhorns. This revelation, of course, spurred my imagination for a Bonanza of gold and I immediately passed this information onto my wife Fran (a very good and dedicated gold dredger), whom was very skeptical of this potential Glory Hole. She was so much unimpressed with my theory, that she wanted no part of a project effort in that area the following season. Well I pointed out all of the common sense, practical and genuine conclusive scientific evidence that supported my theory, but no, we explored another venture that summer. One would think, she didn't have a lot of faith in my judgment, at least when it came to gold mining.

The location where the bucks stopped also had a small alder tree growing behind a large boulder (approximately 12 feet long) in the heart of the drop out area. To me, this illustrates a low pressure zone directly downstream of this large boulder and an ideal area to dredge for gold. As a note of interest, the small alder tree was taken out during high water in 2005 and though, I was unable to interest anyone in dredging this area, it still looked like a great prospect.

The 2007 season dawned with great expectations and I was finally able to convince my wife Fran, that since her dredging time was going to be short due to other commitments, we should try this area. She reluctantly agreed, and we put in and started dredging on the bar side a little downstream and opened up a very nice channel. Results were not the greatest and I'll admit it appeared that the channel had been worked in previous years. Fran hates to dredge in silt and debris. The boulders,though, were huge and had never been moved so we continued on figuring that if they hadn't been worked under, we might uncover something. Fortunately, we found a little gold, far from the Bonanza, I had envisioned but enough to keep Fran interested.

Friends, Gail Scott and Bob Yawn came into camp that week, and had asked me to open up an area, but just leave the gold on bed rock for them to pick up. Now, sure I'm just gonna do that very thing. Wrong!! Fran and I did vacate the channel and jumped over to the downstream side of the boulder where the bucks stopped and punched a hole down to bedrock.

We first had to move Big Momma (the rock), and it was a great deal larger(12ft.) than anticipated, but opened up a good piece of real estate when out of the way. Moving boulders of this size is a great challenge and sometimes is more satisfying than finding a piece of gold. Once moved, it was on with the quest for gold.

Bob and Gail agreed to dredge towards us and we towards them on bed rock. This effort designed to locate a downriver pay streak, if there was one. Twelve days later, following a hamstring mishap to Bob and skin rash to Fran, the movement of two large white boulders uncovered a payday for Bob and Gail. There laying on bed rock, just waiting to be picked up, a veritable handful of jewelry store nuggets. A super finding for a couple of beginning dredgers especially when looking through four feet of water, that much gold looked like a fortune and will remain so for a lifetime of great memories. One of those nuggets made the Roaring Camp Register, a much desired and highly sought after degree of mining recognition. Congratulations to them both on a wonderful and rewarding vacation.

It's always a great pleasure to be part of someone else's success and share their happiness and now the prospects ahead really looked good. We knew the location of the pay streak and all we had to do was open it up and follow it up or down stream. Fran, immediately started picking up some nice nuggets and things were looking great until I attempted to put a ride on a young horse that didn't share my enthusiasm for horseback riding. I more than sorta got bucked off. Fortunately I mostly landed on my head, so the damage to my right shoulder was only bad, not severe. No broken bones, but put outta shape for doing much in the line of dredging.

My contributions were restricted to operation of the winch for rock moving and assisting in breaking plug ups in the dredge suction power jet. This kinda left Fran with all the extra heavy work and as a result sorta slowed our operation down considerable.

On Thursday following my horse mishap, Fran had a nasty rock plug up about five feet from the inlet of her dredge suction hose, and with no help available from me, she walked out of her hole to a place where she could dump the hose out of water. Now here's a lesson in dredging (Please Note) She went to a large flat surfaced rock to dump the blockage (Because she was dredging bedrock in a hot producing area) It was possible there could be gold hung up in that blockage and if so she didn't want to lose it. Guess what? A large nugget rolled out of the hose onto that flat rock that wasn't so flat after all. It slid down the rock headed for a debris pile of hand rocks we had thrown in that area. Fran grabbed for the nugget with a gloved hand and thought she had it. Wrong!! It was gone and I didn't even get to see it! Her comments concerning this loss were some what less than lady like. Sounded more like some old gold miner!!

Things were wrapped up for the day and we didn't get back to dredging until Sunday afternoon. Fran had convinced me that this piece of gold she had lost was worth moving all those rocks we had piled up. Well, we slowly walked out into the river and I was definitely not looking forward to the job ahead. The sun was behind us and ten feet from that not so flat rock and pile underwater, We saw it!! There laying on a hand rock about one foot down and shining like a double eagle gold piece, was our getaway nugget. Very, very carefully Fran eased up to that bunch of loose rocks and picked it up. You know, we didn't work that day, finding that nugget was plenty good enough. There was a lot more gold in that hole to be taken, but circumstances and family commitments made this Fran's last day dredging for the 07 season. Her take for the season, a hand full of beautiful jewelry size gold, had to be enough. Not too shabby for a few days dredging, a bunch of rock moving and all the fun of working at something you really enjoy! Oh Yes! The Gold wasn't too bad either.

Well it was only late July, and the gold season was still in full swing, I had to find another partner to see if that hole was still a good producer. A good friend Don Miller, with family; came to camp and yes he agreed to partner up and help me move rocks and clean it out. Boy! Did he ever work my tail off. He did say, however that he thought we'd lost our marbles, dredging in silt and sticks and leaves, something he's been trained to avoid for twenty years. Then he exclaimed "What's that yellow stuff down there on bedrock??" You know, I told him Fran had said the same thing and wasn't too keen about this area herself until that yellow stuff showed. Aside from his mistrust of my judgment, he's about the best rock rigger, worker, and partner I've ever tied into.

Yes! we did move some rocks, and Yes at the end of his stay, he took home close to an ounce of those beautiful yellow gold pieces that were just laying on bedrock waiting for some grubby ol' gold miner to pick up. Two of those nuggets came in over 2.5 dwt. which put Don on the illustrious Roaring Camp Register twice this year. The little stuff, he wasn't even interested in. Hey Don! How does it feel to be known as a Grubby ol' Gold miner?

Sorry to say, but it's the end of another wonderful Gold Mining Adventure, a great vacation with ol' friends, a lot of excitement, and Fran with another hand full of Gold.

"Sometimes There's a 'REAL' Nice Payday!!"

I've gotta say that in ending this recount of my gold mining experiences, it's been a great pleasure to share with you these adventures, learning experiences, gadgets, tales, stories and to pass on to you, the virtual world class pleasures I've enjoyed for the past 27 years with all my friends and my competition with Mama Nature. This story could go on forever and I hope it does. I'm not wishing to cut out on you in any way but, sun down is getting here and it's time to write "finis" for now. Look for a great tomorrow because I have to say—"As long as I'm a kicking and a diggin, you can bet, That ol'Gold Rush, She Ain't Over Yet!!"

The End

Also available from PublishAmerica

THE UNALTERED CANVAS

by Vanessa Wheeler

In the yard of a man was a tragic murder. One that Jolie Summers had filmed like a true reporter. With her eyes firmly set on the throne of God she wiped her tears and began to embark on a life as an artist. But eyes spied her mad artist skills and took her life and turned it upside down. Could she gain her freedom? Or would she believe in their lives?

Paperback, 108 pages
5.5" x 8.5"
ISBN 1-60672-395-2

About the author:

The author, Vanessa Wheeler, was born in Fresno, California. She is currently working on her second novel. She loves her dogs, Christianity, art, politics, science, and psychology. She is a born-again Christian and hopes her characters reflect Christian imperfections to honor their bravery as believers.

Also available from PublishAmerica

SWIFT EAGLE'S ODYSSEY WITH THE BUFFALO

by Ken Eichler

Swift Eagle's Odyssey with the Buffalo is a sensitive story about a young Indian boy and a huge buffalo that saved the boy from a near drowning in a storm-swollen river and from some hungry wolves. The boy's parents sought help from the Great Spirit to save their boy from harm. Was it by chance that the buffalo was at the right place at the right time, or did the Great Spirit intervene to help the boy? While with the buffalo, Swift Eagle takes a sleep-induced magical journey to the domain of the Great Spirit. The buffalo in the story is from a herd of thousands that were near Swift Eagle's Cheyenne village. The boy's love for the buffalo is heartwarming as is his affection for his playmate Little Flower. You will feel you are at Swift Eagle's homecoming celebration.

Paperback, 59 pages
6" x 9"
ISBN 1-60474-199-6

About the author:

After graduating from high school, I enlisted in the US Navy and served four years (1948-1952). After my discharge from the Navy, I enrolled in the University of Southern California and graduated from the School of Engineering in 1957. I was employed by North American Aviation, Inc. and later by McDonnell Douglas Company (MDC). My principal assignments were in contract administration. I retired from MDC in 1995.

Also available from PublishAmerica

THE END OF SUMMER

by J. Nolting

The End of Summer is a provocative novel filled with excitement, yet it possesses warmth as congenial as a little old lady with frail fingers, adjusting her red hat and admiring the beautiful autumn countryside of the Midwest from the bus as it passes by her...

Amber, a teenage runaway, gets in all kinds of trouble in the red light district of New York City as she seeks fame and fortune as a professional dancer. On her way back home after five years presumed dead, she makes friends with a little old lady, who relates to her that when she was young, she also had a tumultuous past as an Army nurse in WWII.

The two become close friends and exchange the stories of their pasts.

Paperback, 190 pages
6″ x 9″
ISBN 1-4241-3803-5

About the author:

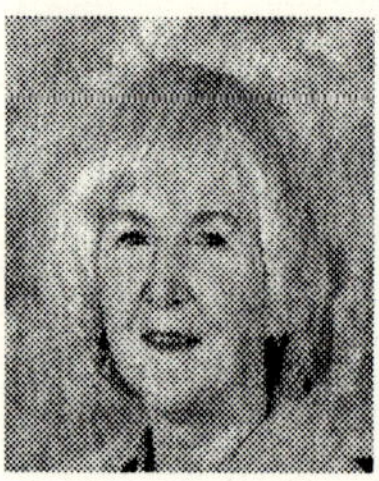

J. Nolting resides in Cedar Falls with her husband. She has two grown children who live out of state. She attended the University of Iowa's writer's summer workshop, the Hearst Center for the Arts, and Hawkeye Community College. In the summers, she tours Iowa and parts of the Midwest on her motorcycle, getting inspiration for her novels. Currently she is working on a third novel.

Also available from PublishAmerica

WEREWOLF ISLAND

by Tony Gardner

Werewolf Island is Tony Gardner's breakthrough novel. He has gained his extraordinary sense of the macabre through his experience running his own haunted attraction. When he is not writing or working, he enjoys flying, scuba diving, fishing, boating and martial arts.

Paperback, 180 pages
5.5" x 8.5"
ISBN 1-4241-6995-X

About the author:

Werewolf Island is Tony Gardner's breakthrough novel. He has gained his extraordinary sense of the macabre through his experience running his own haunted attraction. When he is not writing or working, he enjoys flying, scuba diving, fishing, boating and martial arts.